Les tendances nouvelles de l'ecclésiologie

The Relevance of Physics

Brain, Mind, and Computers
(Lecomte du Noüy Prize, 1970)

The Paradox of Olbers' Paradox

The Milky Way: An Elusive Road for Science

Science and Creation: From Eternal Cycles to an Oscillating Universe

Planets and Planetarians: A History of Theories of the Origin
of Planetary Systems

The Road of Science and the Ways to God
(Gifford Lectures, Edinburgh, 1975 and 1976)

The Origin of Science and the Science of Its Origin
(Freemantle Lectures, Oxford, 1977)

And on This Rock: The Witness of One Land and Two Covenants

Cosmos and Creator

Angels, Apes, and Men

Uneasy Genius: The Life and Work of Pierre Duhem

Lord Gifford and His Lectures: A Centenary Retrospect

The Keys of the Kingdom: A Tool's Witness to Truth

Chance or Reality and Other Essays

* * *

Translations with introduction and notes:

The Ash Wednesday Supper *(Giordano Bruno)*

Cosmological Letters on the Arrangement of the World-Edifice
(J. H. Lambert)

Universal Natural History and Theory of the Heavens *(I. Kant)*

CHESTERTON, A SEER OF SCIENCE

CHESTERTON, A SEER OF SCIENCE

STANLEY L. JAKI

UNIVERSITY OF ILLINOIS PRESS

Urbana and Chicago

This book is printed on acid-free paper

LIBRARY OF CONGRESS CATALOGING-IN-PUBLICATION DATA

Jaki, Stanley L.
  Chesterton, a seer of science.

  Based on lectures presented at the University of Notre
Dame, Apr. 12–15, 1983.
  Bibliography: p.
  Includes index.
  1. Chesterton, G. K. (Gilbert Keith), 1874–1936—Views
on science. 2. Science—Philosophy. 3. Science —History.
I. Title.
Q143.C47J35 1986   500   86-1341
ISBN 0-252-01283-6 (alk. paper)

Contents

Preface

The four chapters of this book are the lectures I gave at the University of Notre Dame April 12–15, 1983. During those days Chestertonians, few of whom had heard of me before, repeatedly asked me about the origin of my interest in their hero. My first encounter with Chesterton was around 1951, shortly after my arrival in the United States. Somebody gave me a copy of *Orthodoxy*, in which I got hardly beyond the second chapter. It contains the remark that unlike poets, mathematicians and cashiers usually go mad, and this was the substance of my familiarity with Chesterton for the next twenty or so years. That around 1973 Chesterton started acting on me as an irresistible magnet has to do with my spotting somewhere a statement of his about the divine strategy that based the Church's superhuman strength on Peter's miserable human weakness. Eager to use that statement in a book I was then writing on the biblical background of Peter's primacy, I had to trace the statement to *Heretics* and study its entire context. From the first chapter it was all too clear to me that Chesterton must be a gold mine for a philosopher and

ix

historian of science. Later it was my good fortune to meet Dr. Russell Kirk as a co-speaker at a symposium on "Freedom, Order, and the University" that Pepperdine University organized for its faculty October 9–11, 1981. Our conversation soon turned to Chesterton and to my steadily strengthening resolve to study Chesterton's reflections on science as soon as my other projects and commitments would be out of the way. This came in the winter of 1982–83. Dr. Kirk quickly secured generous support from the Marguerite Eyer Wilbur Foundation, and Professors Michael Crowe, Frederick Crosson, and other esteemed friends at the University of Notre Dame found several departments and institutes eager to sponsor a series of four lectures. No less valuable was the enthusiasm of Anton Masin, Curator of Rare Books at Notre Dame's Memorial Library, which contains the finest Chesterton collection in the United States. Writing the lectures would have been a sheer delight but for the painful necessity to be selective with a vast material. It is my hope that Chestertonians will now find one more important reason to be proud of G.K.C.

S.L.J.

CHAPTER ONE

Interpreter of Science

A presentation of Gilbert Keith Chesterton as a seer of science may be one of those ventures which he might not have encouraged with the remark that "if a thing is worth doing, it is worth doing badly."[1] He did not speak of science in his *Autobiography* as one of the major areas of his thinking. Those who knew him most intimately provided no encouragement. Science was hardly visible in that charming collection of his best remarks that his wife selected and published in 1911.[2] He undoubtedly approved of it and may have very likely lent a helping hand. He may have been displeased with more than one viewpoint that set the tone of the first major critical study of his work, written two years earlier by his own brother, Cecil. Yet, nothing indicates that he took issue with the declaration there: "In his later books the words 'science' and 'scientists' are used only as terms of flippant abuse."[3] Cecil Chesterton may have in particular thought of the line, "science announced nonentity and art admired decay," in the poem introducing *The Man Who Was Thursday*. Seven years earlier, in 1901, Chesterton had written that "science has given us a vulgar familiarity with the earth, a familiarity without knowledge.[4] In the following years he specified the chief use of science as "to provide long words to cover the errors of the rich,"[5] and extolled philosophy and theology as the fields whose students "may laugh at the rag-tag and bobtail of the science."[6]

Had Chesterton offered only such abusive remarks about science, this would not necessarily mean that this study would be but the airing of his intellectually unrespectable linen. Abusive criticism of a subject may contain insights not to be found in its appraisals, especially if these, as happens all too often, are mere flatteries. But Cecil Chesterton's reference to his brother's "flippant abuse" of science opens a chapter dealing with his assault on the moderns, among whom one would look in vain for mention of a single scientist. With the exception of H. G. Wells, those moderns are not even thinkers preoccupied with science.[7] Thus the possibility arises that Chesterton's abuse of science had for its target not science but some scientifically illiterate moderns' caricature of it, a topic hardly worth studying even if to be found in Chesterton's "latest books." These, by 1909, included *Orthodoxy,* which is unanimously considered the synthesis of his philosophy in its mature form. A potential elimination of the philosophically mature Chesterton as represented in *Orthodoxy* would make even less appealing the study of Chesterton's early years, or a period antedating the first five or so years of Chesterton's career as an author. Chesterton's positive interest in science was assigned by his brother to precisely that period which coincides with the last two decades of the nineteenth century. Those decades certainly fit Cecil Chesterton's statement which is also a characterization of the times: "Nothing in our civilization has seemed to most people more unmistakably creditable than our advance in physical science. Of that advance G.K.C. had in his earlier days spoken with respect—even with enthusiam."[8]

It may therefore seem natural that those who most studied Chesterton's life and thought have not come up with a chapter, not even with a brief section, on Chesterton and science. Maisie Ward, Chesterton's foremost biographer, seemed almost intent on defending her hero against the notion that science had anything to do with him. In quoting an essay in which the sixteen-year-old Chesterton sympathizes with the plight of weavers, Maisie Ward adds: "In this sentiment we recognize the G.K. that is to be, but not when we find him seconding Mr. Bentley in the motion that 'scientific education is much more useful than a classic'."[9] In fact, the motion is most uncharacteristic of schoolboy Chesterton, a less than lackluster student, who half a century later cheerfully recalled having been always at "the bottom of the class."[10] He owed it to his extraordinary ability to do poetry that he did not become a dropout. Writing poems, however good, did not offset his formal scholastic performance to the extent of assuring him membership in St. Paul School's debating society. The motion cited by Ward was made in the "junior" debating society, a wholly private affair run by Chesterton and a handful of friends, none of them much more promising than him academically. According to his own recollection: "It was true that I could never exactly be called a scientific character; and even as between the Classical and Modern sides of my old school, I should always have chosen rather to be idle at Greek than to idle at Chemistry." As to the science which he heard from Uncle Grosjean, a chemist, it did not impress him at all: "It was the same scientific uncle who told me various fairy-tales of fairyland. Thus he told me that when I jumped off a chair, the earth jumped towards me. At that time I took it for granted that this was a lie; or at any rate a joke." Many years later he was still chuckling as he further reminisced: "What Einstein has done with it now is another story—or perhaps another joke."[11] A dozen or so years before this autobiographical revelation saw print, Chesterton went almost naturally from speaking of his ignorance in music "to the more controversial matter of my ignorance in science."[12]

Young Chesterton was not even the kind of modestly systematic or disciplined student who at least in Greek and Latin, or in the humanities in general, would have had a chance in Oxford, let alone in Cambridge, where by osmosis he might have learned science, a subject certainly not treated at Slade School of Art, where he finished his formal education. Typically, the techniques of drawing did not appeal to him in the slightest.[13] Disciplined study, be it of art, was not Chesterton's forte. While officially enrolled in Slade School of Art, a branch of the University of London, he liked to sit in classes in other departments. This is a fact that Chesterton himself may have disclosed in 1927 to the Provost of the University, Sir Gregory Foster, who introduced him as the seventh of distinguished lecturers invited to help celebrate the University's centenary.[14]

4

Whether Chesterton did in fact attend some science classes is, of course, a matter of conjecture. Yet, he himself stressed that he grew up in times when it was impossible not to take note of science: "Science," he recalled, "was in the air of all that Victorian world, and children and boys were affected by the picturesque aspects of it."[15] About a mind like Chesterton's, so attached to images and so great a master in conjuring up their startling varieties, this recollection may be more significant than its brevity may suggest. The image of science which sprang up in his mind was as vivid and large as the Crystal Palace, which he visited on more than one occasion. It was the symbol of the cosmos for the man of his youth. Not that he was wholly taken by the Crystal Palace and the world of science it represented; he found about it "something negative. . . . arching over all our heads, a roof as remote as a sky . . . impartial and impersonal. . . . Our attention was fixed on the exhibits, which were all carefully ticketed and arranged in rows; for it was the age of science."[16]

The unmistakably disparaging ring of the remark, "age of science," may have been more than enough to prevent students of Chesterton from seeing any promise in a topic like Chesterton and science. None of them found promising in that respect Chesterton's remark made almost in the same breath that while he did not love such big scientific instruments as a telescope, he loved small scientific things, such as a microscope.[17] Promise is, however, also an expectation. It was not to be expected that the subject should seem promising at all to Chesterton scholars. They all belong to that world of letters which, rightly or wrongly, has, for the past two decades, been ever more loudly denounced for its insensitivity to a vast new world around it, the world of science. Without prejudging the matter, let alone endorsing C. P. Snow's bland claim that scientists, especially the engineers of regimented societies, are less insensitive to the world of letters, or humanities, than humanists are to the sciences,[18] the characterization of all major students of Chesterton as men of letters should be above dispute. Even when they were not primarily literary critics, their chief interest was letters in a broader sense: philosophy, theology, history, social problems and the like. Thus even on the assumption that science is noticeable in the mature Chesterton's thought, they would not be the ones to be attracted to it, let alone perceive and set forth its relevance.

The reference to C. P. Snow is all the more appropriate as he represents the kind of humanist or man of letters who is also strongly rationalist, and therefore with a distinct bias in favor of science. Such people—and most of Chesterton's first critics came from that group—were indeed blinded by his flippancy about science to all else he had said about it. The first of such critics, Julius West, who died at the early age of twenty-seven, in 1918, set a pattern by speaking of Chesterton's "detestation" and "abuse" of science. He did so by insisting on the "miserable failure" of Chesterton's endeavor "to show that the triumph of atheism would lead to the triumph of a callous and inhuman body of scientists," as if nothing else was to be said on the subject.[19] Much the same holds true of Alan Handsacre's *Authordoxy*, in which Chesterton's paradoxes are, one after another, turned against him with undeniable stylistic skill,[20] and of the no less sparkling study by Gerard William Bullett, also from the early 1920s.[21]

Perhaps Chesterton's so-called abuse of science prevented some early students of Chesterton, very much in sympathy with the value of his thought of Christian apologetics and philosophy, to see a gold mine there with respect to a proper understanding of science. The earliest of these students, Father Joseph de Tonquédec, produced a study on Chesterton, which is all the more a disappointment in that respect as the alleged incompatibility of science and religion was a battle cry in the France of his time.[22] The subject of Chesterton and science was just as unexploited by W. F. R. Hardie, for whom Chesterton was a chief defender of Christian humanism. Although Hardie emphatically noted that Chesterton feared a threat to Christianity from two quarters, the "tyranny of science" and the "tyranny of the contemporary," about Chesterton's battling the "tyranny of science" he found worth mentioning only Chesterton's antagonism to medical men promoting eugenics.[23] Chesterton's dicta on science meant nothing to Patrick Braybrooke, who, in addition to a study of Chesterton's thought, published, also in the 1920s, a prose and verse anthology of Chesterton the Catholic and an encomium of Chesterton the Sage.[24]

The relevance of science for Chesterton's thought, and the relevance of it for the interpretation of science, were not broached in studies of him mainly prompted by his death, an omission with particularly frustrating aspects in each case. After all, the Rev. John O'Connor, author of

Father Brown on Chesterton, was among other things "a man of science" in Chesterton's own estimate. It is also in that book that one finds quoted in some detail a review of *Orthodoxy* distinctly hostile to it, because Chesterton allegedly finds "the first postulate of Science, the uniformity of Nature, . . . too dull to believe in."[25] There is no discussion of science in an essay by Maurice Evans which won a coveted prize,[26] or in Frank Alfred Lea's *The Wild Knight of Battersea,*[27] described decades later as perhaps the wisest brief treatment of Chesterton available.[28] Yet, it may not have been really wise on Lea's part not to follow up a lead which he obviously spotted as he remarked: "There seems, at first sight, to be only one province of human activity which he left exclusively to Mr. Belloc, and that was science. He was, as a matter of fact, always inclined to be less than just to both science and scientists." Lea, who then stated that "Chesterton left science to scientists," failed to redeem the promise implied in that "first sight."[29] Not even a second thought, let along a hindsight, was in fact necessary to see the importance of a statement of Chesterton in so prominent a context as his *Orthodoxy* could provide: "I read the scientific and sceptical literature of my time—all of it, at least, that I could find written in English and lying about. . . . It was Huxley, Herbert Spencer, and Bradlaugh who brought me back to orthodox theology."[30] All three swore by science, in particular Charles Bradlaugh, who for six years was kept out of Parliament because he refused to take the oath required from new MPs on the ground that he was an atheist. The England of the 1870s and 1880s saw no professed atheist with

greater impact than Bradlaugh. As one would expect, Bradlaugh, president of the Freethought Society, heavily relied
on science as the factor which, as he put it, "has razed
altar after altar heretofore erected to the unknown gods,
and pulled down deity after deity from the pedestals on
which ignorance and superstition had erected them."[31]

T. H. Huxley and Herbert Spencer, the two others named
by Chesterton as the ones who brought him back to
Christianity, should be familiar figures. They were at one
not only in writing mostly on science, but also in writing
about it in a way which, if not always truly informative
and profound, was invariably attractive. They were also at
one in harnessing science in the service of scientism. Scientism, or the claim that only the scientific or quantitative
method yields valid knowledge and reliable value judge-
ments, provoked Chesterton to many devastating and penetrating remarks, which call for a separate discussion. Here,
in the context of this rapid survey of Chesterton studies,
scientism deserves a moment's reflection, in that the first
major study of Chesterton after the one by Lea was devoted
to the role of paradox, a topic that in itself implied the
very opposite to the claims of scientism. Hugh Kenner in
his study of the role of paradox in Chesterton's literary
artistry[32] rightly argued that behind Chesterton's continual
reliance on paradoxes there lay the ever alive philosophical
conviction of a born realist, indeed of a realist metaphy-
sician. Yet, Kenner did not so much as hint that it is the
restriction to the univocal within science and the universal
validity assigned by scientism to that restriction which
represent the starkest and most timely contrast to the

doctrine of the analogy of being, the cornerstone of realist metaphysics. Such was an omission all the more strange, because many of Chesterton's sparkling paradoxes were meant to vindicate and bring back into broad awareness precisely that doctrine.

One would find little appreciation of Chesterton's dicta on science in the numerous studies that appeared during the last two decades straddling the 1974 centenary of his birth. The few pages that Christopher Hollis devoted at the end of his *The Mind of Chesterton*[33] to his views on evolution were less on Chesterton than on Teilhard de Chardin, who in his *Phenomenon of Man,* to say nothing of his other meditations, was more a man of letters than a man of science. Science appears only in incidental remarks in another lengthy portrayal of Chesterton's thought by Sister M. Carol.[34] The same holds true of Gary Wills's effort to unmask the real wellsprings of Chesterton's intellectual concerns.[35] How close a man of letters could come to the heart of the matter and how far removed from it he could remain is best illustrated in a penetrating essay on Chesterton's thought by Stephen Medcalf, a lecturer in English. In that essay, which would have done credit to such outstanding champions of philosophical realism as a Gilson or a Maritain, Medcalf rightly noted that "Chesterton's inner quality is to communicate knowledge of our capacity for believing in God as creator and of enjoying our position as creatures." If, however, the perspective of all tangible and intangible existence as something created is so central in Chesterton's thought, the question imposes itself: Why not bring in that very science which has revealed so much about tangible reality and claimed all too often competence about its intangible kind as well? After all, was not Medcalf's starting point that very solipsism insofar as it sets the tone of much of twentieth-century thought, scientific thought not excepted? But he overlooked an obvious pointer: Arthur Stanley Eddington, a prominent physicist as well as an unabashed solipsist, was the first illustration in *The Common Sky,* a major study of solipsism in modern literature to whose author Medcalf felt heavily indebted.[36] Medcalf's possible excuse that the mature Chesterton himself did not write about science often and emphatically enough is, as will be seen, a poor excuse for not taking very seriously what he said on science.

About the less than 200 smallish pages of Lawrence J. Clipper's monograph on all major aspects of Chesterton, one can at least say that it is a tour de force of condensation. But Clipper had no right to ignore Chesterton's statements on science if it was true, to quote him, that "if one were to characterize Chesterton in a few words, it would be no great misrepresentation to say that he was a champion of religion against the combined forces of secularism, science and Positivism." One can easily miss the gist of a vast strategy if little or nothing is said about the forces it is aimed at, especially if it is true, as Clipper put it, that Chesterton's "alarm is generated by the endless aggrandizement of science."[37] It was that aggrandizement which Chesterton saw behind the mechanization of life, and in particular behind the growing subjugation of small landowners and artisans to the Goliath of monopolies, agricultural, industrial and technological. To paint a portrait of

Chesterton, the distributist, with no regard to what he said on science, might give credit to a sociologist or an economist, but the portrait will lack Chestertonian depth.[38]

The Chesterton scholars who might, at first sight, escape censure here are the literary critics and at times fellow poets, but their case is less objectionable than it may appear. W.H. Auden would have fared better had he put together a selection from Chesterton's poetry rather than his nonfictional prose. In his introduction to the prose anthology, Auden saw Chesterton at his best in exposing "the hidden dogmas of anthropologists, psychologists and their ilk who claim to be purely objective and scientific."[39] For all that, the selections are very short on what Chesterton supposedly is at his best. Such a remark would, of course, be wide of the mark had Auden had an inkling of what was the true message of by far the longest selection he offered. It was "The Ethics of Elfland," which he took for Chesterton's most enjoyable prose piece because there Chesterton is "at his silliest."[40] Lynette Hunter, another recent and prominent literary critic of Chesterton, perhaps would not disagree, if chapter 6 in her book[41] is taken for the emergence of Chesterton in all his essentials as a literary creator bent as much on a message as on his special way of conveying it. That chapter, aptly called "Developing the Land," covers the years 1908–12 and centers, not unexpectedly, on *Orthodoxy*. Even more important, at the very start of that chapter she invokes chapter 4 of *Orthodoxy* because at its end Chesterton sums up his philosophy in five points. She does not so much as suggest the contents and gist of that chapter, something that seems crucial if it logically leads up to that summary. That chapter, "The Ethics of Elfland," is, in fact, as will be seen shortly, one of the most penetrating discourses on the nature of scientific reasoning that has been so far produced. In Ian Boyd's analysis of Chesterton as a novelist there is at the very start a lack of much needed emphasis on science. Start as he had to with *The Ball and the Cross*, he let its very opening, Professor Lucifer's advocacy of science, slip through his otherwise firm grasp on the subject.[42] The *Outline of Sanity*,[43] the latest and much advertised biography of Chesterton, offers nothing on science, although, as will be seen, proper thinking about science was for him very germane to sanity. The author of that biography, Alzina S. Dale, could not, for all her familiarity with the literature, have been alerted by two brief essays on Chesterton and technology published in 1976.[44] Nor would she have found a single scientist among those many whose reminiscences about Chesterton were put together into a delightful volume by Cyril Clemens shortly after Chesterton's death.[45]

This brief survey of major books and essays on Chesterton's thought would be very incomplete if it contained no reference to two works, both as unpretentious, as immensely useful. One of them is Joseph Sprug's *Index to Chesterton*, published in 1965.[46] There the headings 'science' and 'evolution' contain no fewer entries than 'religion' and 'Catholic Church.' The entries under 'science,' 'science and religion,' and 'scientists,' amount to over two hundred, hardly a negligible quantity. The other work is the Chesterton bibliography published in 1958, with a supplement added in 1968, by John Sullivan.[47] Unlike the *Index,* in which science and related items are rather prominent, the bibliography yields practically nothing as far as science is concerned. The only silver lining in that "no-science" cloud of titles is that a discussion of Chesterton and science may not be altogether unoriginal.

Sullivan's bibliography is noteworthy for a far more important reason. While one is often astonished by his ability to spot material relating to Chesterton in most unlikely sources, no less astonishing may seem his failure to spot what may easily be the most often printed Chesterton text apart from such perennial Chesterton bestsellers as *Orthodoxy, The Everlasting Man, St. Thomas Aquinas* and the Father Brown stories. The text in question is about one-third of chapter 4 of *Orthodoxy,* "The Ethics of Elfland," reprinted in 1957 in, of all places, *Great Essays in Science,* a title in the Pocket Library.[48] A typical first printing of titles in that series was in the tens of thousands, and copies were available not only in all bookshops but also at many newsstands in the 1950s and 1960s. There was Chesterton in the company of Albert Einstein, Charles Darwin, Henri Fabre, J.R. Oppenheimer, Arthur Stanley Eddington, Alfred North Whitehead, and Bertrand Russell, so many giants in mathematics, physics, and natural history. Chesterton was also in the company of such prominent interpreters of science as John Dewey, Ernest Nagel, and even T.H. and Julian Huxley. In such a company Chesterton needed a special introduction if not plain justification. Martin Gardner, who as associate editor of *American Scientist* put together that volume, did indeed apologize: "It may come as a shock to many readers," he began his introduction of Chesterton, "to find a selection by Gilbert Keith Chesterton (1874–1936) included here. The rotund British writer was not noted for his knowledge of things scientific. ... Yet there are times, as in the following selection, when he startles you with unexpected scientific insights." Worse, Gardner noted in way of final forewarning, the selection came "from, of all places, *Orthodoxy*, Chesterton's most famous work of Christian apologetics," a work published, Gardner added, perhaps to take some of the sting out of the whole business, "fourteen years before Chesterton became a Catholic." There was, of course, one unquestionable compensation for being exposed to Christian apologetics at its best. It was Chesterton's style "for which the author is justly famous—brilliant, witty, alliterative, dazzling in its metaphors and verbal swordplay, and a joy to read even when you disagree with him."[49]

The re-christening by Gardner of the selection as "The Logic of Elfland" might have prompted Chesterton to some pointed remarks going far beyond a lecture on editorial ethics. In the whole section quoted, and in fact in the entire chapter, the word logic is hardly to be found. Not that there is no logic in it, but it contains much more. Hence Chesterton's choice of the title, "The Ethics of Elfland," but this is to anticipate. A summing up of the selection is not an easy task, as it is never easy to give a concise and systematic outline of any of Chesterton's philosophical chapters and books. A philosopher of tremendous incisiveness, he is never discursive. As he himself put it in the first page of *Orthodoxy*, he was giving his philosophy "in a vague and personal way, in a set of mental pictures, rather than in a series of deductions." The selection[50] begins with a picturesque stroll into fairyland, the sunny country of common sense, the land of Jack the Giant Killer and Cinderella. Or to hear Chesterton introduce that stroll: "My first and last philosophy, that which I believe in with unbroken certainty, I learnt in the nursery." In that land necessity properly so-called, that is, a necessity which is strictly necessary, is restricted to mathematical or logical sequences and implications. If Ugly Sisters are older than Cinderella, then Cinderella is necessarily younger than Ugly Sisters. Not so beyond the hedges of Elfland. There bespectacled men, scientists, with Newton in the van, take sequences of events as necessary and call them laws. They would be entitled to call necessary, that is, a law, only a logical reciprocity, such as if Newton's head was hit by an apple, then the apple was hit by Newton's head. Yet, they call a law, that is, a necessity, any fall of any apple, although they witness only "weird" repetitions.[51] In fairyland muddled thinking arises only when the blow of a horn is taken for the necessary cause of the collapse of the enchanted castle. In science-land, heads are customarily muddled through the habit of attributing a necessary mental connection between the apple's leaving the branch and the apple's reaching the ground. Chesterton then declares in the style of a true philosopher, a style which he hardly ever allows to himself for more that a few lines: "A law implies that we knew the nature of generalization and enactment; not merely that we have noticed some of the effects." Of course, and this cannot be emphasized enough, by law Chesterton once more means strict necessity, a meaning which strangely enough is easily lost from sight in this scientific age of scientific laws. "All the terms used in the science books—'law,' 'necessity,' 'order,' 'tendency,' and so on, are really unintellectual because they assume an inner synthesis which we do not possess." Chesterton was aiming at T. H. Huxley,[52] who saw an argument for unalterable law in a count of the ordinary sequence of events. Extensive as any such count was, it could not be strictly conclusive: "We do not count on it; we bet on it. We risk the remote possibility of a miracle."

Chesterton's use of the word miracle in this context is most significant. On a cursory look he merely restated an old observation often tied to David Hume that inductions, however widely based on counting facts, never become conclusive in the sense of a logical necessity. But for Hume the lack of necessity in induction was a basis on which to argue against miracles as well. Such was the voice of a skeptic trying to secure universal certainty to skepticism. Not much different was the jeer with which Bertrand Russell rebounded from the threshold of realist metaphysics, when he spotted it lurking in the background as the sole justification of any inductive conclusion, be it most scientific. If, after having been nourished by uncounted meals, the absolute certainty that his next meal would nourish him was a condition that he should take it, he would, he wrote, rather renounce rationality than miss that meal.[53] No wonder that Bertrand Russell, the philosopher, came in for Chesterton's stricture, though in another context.[54] For in Chesterton's eyes the whole lesson to be gained from a reflection, however elementary and available already in the nursery, on the status of induction and on the whole status of scientific laws, was a lesson in true metaphysics.

In a truly Aristotelian fashion Chesterton (who in the "Ethics of Elfland" merely spoke of the "ancient instinct of astonishment" and endorsed elsewhere Aristotle's $\mu\acute{\epsilon}\sigma\sigma\nu$ or philosophical middle road)[55] begins with wonderment at reality. It is that ability to wonder at reality which is emphasized and upheld in the rest of the section quoted by Gardner. To have that wonder one needs to overcome the sentimentalism of the scientific man who is "soaked up and swept away by mere associations" and who binds moon and tides into a necessary unity, just because he sees them together without seeing through what he is doing. Rather, one has to assume the stance of the cool rationalist from fairyland who does not see why, in the abstract, that is, on a strictly logical ground, "the apple tree should not grow crimson tulips. It sometimes does in his country." In other words, a choice is to be made: One is either to continue in the deadly routine of forgetting that all the so-called rationality rests on wonderment, or one is to break the continuity of that routine by remembering occasionally that one is in the deadly habit of forgetting.

My excuse for paraphrasing Chesterton here is that in this case his paradox is unusually rich in concise allusions. They had to be somewhat streamlined if developed and written in the old discursive style of "learned" essays. Another Chesterton alone could render his thought in striking paradoxes the use of which he once stated "is to awaken the mind."[56] That this pregnant definition of the purpose of Chesterton's favorite device is from his most philosophical book, *St. Thomas Aquinas,* is more than a mere coincidence. There metaphysical realism obtained an articulation of which unexpected profundity was no less a mark than gripping vividness. To both no less a master of Thomist realism and of graceful style than Etienne Gilson gave supreme homage. Gilson wrote in 1933 that he felt so much bested by Chesterton's book as to become momentarily discouraged about writing anything more on Thomas and Thomism.[57] Four years earlier, Gilson had heard Chesterton's lectures at the University of Toronto—the greatest intellectual revelation of his life, he was to write in a 1966 letter. Gilson was astonished at Chesterton's ability to anchor his starting point invariably and with unfailing ease in the intellectually perceived reality.[58] Such a start was the very reverse of the procedure of most philosophers since Descartes and Kant, who begin with ideas and, as all the history of modern philosophy shows, never get to reality. It should therefore be no surprise that Martin Gardner, who used, if not abused, Chesterton as a witness on behalf of a notion of science that is purely operationist in the counter-metaphysical sense, had no use whatever for the remainder of "The Ethics of Elfland." The remainder was in fact the remaining two-thirds of a chapter, in which Chesterton, in full witness to the philosopher he was, gave a foretaste of the philosophically deepest pages of his *St. Thomas Aquinas.*

From the wonder of a sudden remembrance that reality is incomparably more than what can be accommodated in the cubbyholes of logical sequences, there is but one step to what Chesterton called the "second great principle of fairy philosophy,"[59] the realization that reality both in its smallest details and in its vast entirety was most specific. The specificity or queerness of things, both when taken singly and when taken together, that is, as a cosmos or universe, was in Chesterton's judgement the means par excellence to restore sensitivity for the real. Such sensitivity was a commitment to a reality much more than sheer logic, and this is the reason Chesterton spoke not of the logic but of the ethics of elfland. Consideration of specificity was also the means of making it dawn on the onlooker that a ubiquitous and all-embracing specificity is the result of a choice transcending the cosmos, a choice which is therefore the sole privilege of a Creator. Such is in a nutshell Chesterton's cosmology, which calls for a separate discussion.

One aspect of that cosmology should be mentioned here, however, because it is closely related to Chesterton's immediate concern about the proper interpretation of scientific law. He rightly perceived that the vast, if not infinite, mechanistic universe as celebrated by Spencer, Haeckel, and their ilk, had for its real purpose the destruction of man as a metaphysical being. The vastness of that universe made man look puny and insignificant, whereas the laws of science conceived in terms of a mechanistic ontology pre-empted man's freedom and the possibility of any genuine novelty in the universe. Once the ontological emptiness of the notion of scientific law was exposed and a realist grasp of existence achieved, reality could be seen as the product of a superior will and therefore a place germane to the exercise of free human will. Repetitions in nature were then less the monotony of a clockwork than so many "theatrical encores." The laws of science issuing in scientific fatalism had therefore to appear just as groundless as the assertions of the great determinists of the nineteenth century, who somehow seemed to undermine this very feeling that "something had *happened* an instant before or whether in fact anything *happened* since the beginning of the world" (italics added).[60]

Written in 1908, "The Ethics of Elfland," with its heroic and profound assertion of novelty, reminds one of Bergson's *Evolution créatrice*. The latter, first published in 1906, was greeted by many at that time as the long awaited liberation from the ironclad mechanism of science. The mystical and at times mystery-mongering *élan vital*, on the basis of which Bergson tried to vindicate novelty and freedom, has long since lost its luster and convincingness. Though Chesterton was accused of intellectual reactionarism and provincialism for having ignored many of his contemporary luminaries, Bergson among others,[61] he was not ignorant of an incomparably more reliable assurance of novelty and all that goes with it, freedom and the rest. That assurance was realist metaphysics, which alone saves science from turning into a game that quickly turns not only against its practitioners, but against mankind itself. That assurance could not have come to Chesterton, say, from Karl Pearson's *The Grammar of Science*. Chesterton, whose first reference to Pearson dates from around 1912 and in the rather different context of eugenics, may very well have seen Pearson's book before writing *Orthodoxy;* the book was in great vogue in England from the moment of its publication in 1892.[62] In a true positivist fashion Pearson demythologized scientific law of all ontological connotation. He did so on the basis of the positivist myth that man's grasp of reality was nothing short of a "metaphysical" illusion. Much the same was the message of other exponents of scientific positivism, such as Poincaré and Mach. The latter's positivist account of the history of mechanics would have been too technical for Chesterton. The former was at his best with paradoxes, such as, for instance, that any determinist argues in a non-determinist fashion, which did not best the paradoxes of Chesterton. Nor did Chesterton need to read Poincaré to be startled by the solipsism which positivism generates and to which Poincaré subscribed with his dictum that only thought exists.[63] By 1908, when he was thirty-four, Chesterton's wrestling himself free from the lethal grip of solipsism was a sixteen-year-old story.[64] It did not indicate morbidity, but rather revealed the precious genius which led him, before he was twenty, to see vividly the most fundamental issue in philosophy. It was an intellectual drama, in which the agony of facing up to a possible disintegration of coherence was followed by the ecstasy of the assurance of real things (they alone could, as Belloc aptly noted, satiate his mental hunger[65])—a drama that made a deep imprint on him for the rest of his life. Spirited opponent as he became of materialism, it was for him a far lesser evil than solipsism, be it euphemistically called idealism. In *The Poet and the Lunatics*, subtitled "Episodes in the Life of Gabriel Gale," Gale's outburst aimed with a galewind's force at an apologetic materialist, is Chesterton speaking: "Afraid!" cried Gale as if with indignation; "*afraid* you are a materialist! You haven't got much notion of what there really is to be afraid of! Materialists are all right; they are at least near enough to heaven to accept the earth and not imagine they made it. The dreadful doubts are not the doubts of the materialist. The dreadful doubts, the deadly and damnable doubts, are the doubts of the idealist."[66]

One wonders what Chesterton's reaction would have been had he read Pierre Duhem, the only positivist philosopher of science who, while stressing the purely formalistic character of scientific laws, held high that common sense, so dear to Chesterton, which alone could secure, by its grasp of reality, meaning to science.[67] Duhem's *Théorie physique* was published in 1906, two years before the *Orthodoxy,* but it came out in English translation only in 1954 at a time when logical positivism was at its overweening zenith in England as well as in the United States. Typically, every effort was made to turn Duhem into a positivist, as if he had never asserted the primacy of objective reality, to which common sense and not the exertions of logic, scientific or not, gave the only access.[68] Such is the background of Gardner's tactic in cutting off "The Ethics of Elfland" at a point where it transcends logic and becomes a commitment to reality, that is, an epistemological ethos.

The quarter of a century that has since elapsed was long enough to let the inner logic of logical positivism play havoc with itself and especially with its most cherished field, the interpretation of science. Logical positivists were the first to sound alarm when there appeared Thomas S. Kuhn's *The Structure of Scientific Revolutions,* a book whose principal instructiveness will forever lie in its author's consistency. Since logic, in itself a purely private matter, must thrive on something, the most promising and logically allowable terrain for it had to be that psychologism which hardly leads its cultivator beyond the self in particular and solipsism in general. Thus when Kuhn put science into the straitjacket of psychological mutations, science immediately fragmented into links forming no chain, or rather a mere sequence of revolutions. There no room was left for coherent progress, the traditional and hallowed distinction which, in the eyes of erstwhile positivists, science alone possessed. Such was the background of those traumatic and futile disputes that held in their grip philosophers and historians of science in the late sixties and through much of the seventies.[69] The flood of ink produced by those debates only illustrated the truth of a largely forgotten remark of Chesterton in "The Suicide of Thought," a chapter which immediately precedes "The Ethics of Elfland." The remark is toward the end of that chapter, which completes, in Chesterton's own words, "the first and dullest business of this book—the rough review of recent thought." The relevance of that remark to present-day debates tearing science to pieces derives from its truth which, if such, is universally applicable and forever valid. What Chesterton does in the context is to denounce the shift from the objectivity of true liberalism into the carefree subjectivity of the "liberals'" liberality, that is, their endless programs of liberalizations, reforms, and revolutions, programs that include no place for objective constraints on liberty. "Men," Chesterton sums up the process, "have tried to turn 'revolutionize' from a transitive to an intransitive verb."[70]

When a phrase is so concise as to admit no improvement, a brief reflection on it is never superfluous. To revolutionize in an intransitive sense is to cavort in change for the sake of change, which is the very opposite of true liberalism, for which liberty is never something subordinate to revolutions. As always, here too there is a deeply philosophical side to the matter. To revolutionize in an intransitive sense is a studied disregard for the thing or things changing, and also a contempt for the concern whether anything remains identical in the change, let alone whether change truly means progress, that is, an increase of objective value content. Such a reflection should evoke Aristotle, who based all his philosophy on viewing change neither as illusory nor as radical, that is, to use the modern term, revolutionary. He was a champion of continuity through change, which comes through even in his political theory. There he made it a point that dictatorships or tyrannies can only be overthrown from the outside.[71] Indeed, only a move from the outside, a thrust from objective reality, perhaps in the form of a major economic crisis, would ever overthrow the academic dictatorship whose trendsetters cavort in scientific and above all in ethical revolutions. Herein lies the principal and lasting relevance of Chesterton for science which dominates modern thought even more than was the case in his day. Since then no one has argued more persuasively on behalf of objective reality as a safeguard of sanity, including the sanity of science.

Chesterton's stunning insistence in "The Ethics of Elfland," that science as such gives only logical identities and relations but no realities, should make him appear an interpreter of science to be ranked with a Duhem and a Meyerson. The latter, it is well to recall, published his *Identité et réalité* in 1907, only a year before "The Ethics of Elfland" saw print. But Emile Meyerson dared to spell out only in a muted and at times very roundabout way the inference that if only the identity relations of mathematical physics existed, nothing in reality would ever *happen*.[72] Chesterton never beat around the bush, in fact around any bush or reality, let alone around the supreme reality which is the burning bush. He was far ahead of his times even as an interpreter of science, precisely because he was a respecter of that reality which transcends the moment. He could not suspect, when he died in 1936, that the academic scene was ready to recast science into a series of intellectual mutations for the sake of mutations,[73] but his great phrase "to revolutionize in an intransitive sense" was ready for those who looked for a powerful antidote against the new epidemic in "learnedness."

Chesterton was already dead when the bearing of his defense of reality on the sanity of science received an unexpected endorsement from Einstein himself. Not that Einstein ever read Chesterton. He might not even have heard his name. Yet, Chesterton's concern for reality should seem to have been given an unintended vote of support in what became one of Einstein's last and agonizing remarks on the most agitated philosophical issue prompted by modern physics. The issue is the coherence or incoherence of reality, and in fact the issue of whether there is an objective reality or all reality is the product of the mind. According to Einstein, the great majority of modern physicists failed to realize "what a risky game they played with reality."[74] They did so because they sided with the so-called Copenhagen school of quantum mechanics, a school more of philosophy than of physics. According to that school, all processes in nature are rooted in chance—a word which, it must be noted, is never rigorously defined by the proponents of that philosophy. Their principal argument is Heisenberg's indeterminacy principle, according to which it is impossible to measure with full precision, that is, with no margin of error, certain processes on the atomic level. But in their philosophy this argument means to say that a process that cannot be measured with complete precision cannot take place with a complete causal exactitude. About this *non sequitur,* or simple rape of pure and innocent logic, which has been swallowed by such an overwhelming majority of twentieth-century scientists and "scientific" philosophers so as to become a climate of opinion,[75] still by far the most expressive indictment is a statement of Chesterton: "Modern science cares far less for pure logic than a dancing dervish."[76] The presence and popularity of such dancers is hardly a sign of sanity. It would be in greater supply in this scientific age had attention been paid to Chesterton, the interpreter of science.

CHAPTER TWO

Antagonist of Scientism

As with a man, so with a trend. It has to exist before it is given a name. Scientism had for long been a trend, before it emerged as a label for an intellectual imperialism that makes ever greater demands in the name of science. In the England of Chesterton's youth and early manhood, scientism had voluble and overbearing spokesmen such as Herbert Spencer, Ernst Haeckel, and H. G. Wells. A recall of Spencer is all the more justified, given that in "The Ethics of Elfland" (though not in the part selected by Gardner for his *Great Essays in Science,* as discussed in the previous chapter) Chesterton chastized Spencer as a chief representative of an ideological imperialism of the lowest type."[1] The hold which Spencer had on Anglo-Saxon thought in the closing decades of the nineteenth century cannot be exaggerated.[2] Though he held court as an undisputed interpreter of science, in no branch of it did he have a systematic training. Almost from the start of his parlaying himself into an oracle of his age he kept referring to science as the highest tribunal. His voluminous and bewitchingly facile writings were but a commentary on a long essay he wrote in 1860.[3] There he ascribed supreme competency to science in all fields, which he passed in review as a dictator by raising again and again the question, "What knowledge is of most worth?" And whether he addressed the question to education, to national life, to the arts, to religion, to character building, or to family care, the invariable answer was Science. Scientism could not have received a more sweeping manifesto.

In the context just referred to, Chesterton was wise enough to take an aim at Spencerian scientism on a specific point. General attacks, he knew all too well, readily dissipate themselves in mere generalities. The point was the extension of a particular scientific finding to far beyond its purely quantitative relevance. The extension smacked of imperialism not only because of its authoritative tone. The slighting of man in a philosophical sense, on account of his quantitative puniness compared with planetary dimensions, reminded Chesterton of the contemptuous slighting by British imperialists of the Irish on account of the relative smallness of their number and territory.[4] This intellectual imperialism was recalled by Chesterton as a chief characteristic of the times in which he grew up: "I have heard through my youth and early manhood, the repetition of the ultimatum: 'You must accept the conclusions of science.'"[5]

Concerning particulars, Chesterton's chief target was not Spencer, who wrote more and more philosophy and read less and less science as time went on. The science book which in the England of Chesterton's youth and early manhood served as a sacred manual of rationalist creed was an import to England, Haeckel's *Schöpfungsgeschichte*, which in its English translation, *History of Creation,* saw every year a new edition or at least a new printing between 1876 and 1901. Haeckel's book was an ideological import only in the sense that he spelled out so clearly and so fully the implications of Darwin's theory that, on seeing the original German, Darwin hesitated about pressing on with his *Descent of Man.*[6] It was with that book that Haeckel became the English popularizer of all science, from the genesis of galaxies to the origin of life, although England had such brilliant popularizers of science as T. H. Huxley, Richard Anthony Proctor, John Ellard Gore, and others. Haeckel was intent on far more than science. "All that the layman heard of it [science]," Chesterton recalled, "was dominated by the now dead materialism of Haeckel."[7] Worse, social pressure demanded its acceptance. Although the author of the *History of Creation* was clearly "less of a scientist than a propagandist, and a pretty unscrupulous propagandist," Chesterton reminisced in another context, "we were all supposed to swallow what he said at once, because he was Science."[8]

The early manhood of Chesterton included the six years that witnessed, between 1895 and 1901, the publication by H. G. Wells, Chesterton's senior by eight years, of books that inflamed if not the mind at least the imagination of even the scientifically illiterate. *The Time Machine, The Island of Dr. Moreau, The Wheels of Chance, The Invisible Man, The War of the Worlds,* and *The First Man on the Moon* carried far and wide, and in intoxicating doses, the message that through science everything became possible. Through those books "the shape of things to come" became a cultural war cry.[9] There was much more than science fiction coated with sentimentalism to H. G. Wells's science romances, the first of which, *The Time Machine*, sold 6000 copies in a few weeks. Nor was Wells a mere English Jules Verne who, unlike Wells, was careful with scientific details. Wells aimed above all at conveying the fearsome message of T. H. Huxley, whose biology student he was, without retaining that illogical partial retraction through which Huxley wanted to take some of the sting out of that message. While Huxley kept referring to an ethical revolution issuing in brotherhood, which for a while might keep in check the cruelty of biological struggle,[10] Wells had no use for such heroic *non sequiturs*. He portrayed so forcefully the grimness of evolution that a critic of *The Island of Dr. Moreau* wondered: "The horrors described by Mr. Wells in his latest book very pertinently raise the question how far it is legitimate to create feelings of disgust in a work of art."[11] The quintessence of the book was, according to Wells's own recollection, "the hideous grimace in which the universe projects itself."[12] Whatever Chesterton's admiration for Wells's sincerity and courage in setting forth the pessimism of Darwinian materialism, he was not inclined to see in Wells's technique a useful way of eliciting the spark of optimism through the shock of pessimism.[13] Indeed, cosmic pessimism was to claim Wells himself as one of its notable modern victims.[14] Chesterton had no use whatever for a flirtation with pessimism that, as he rightly sensed, exuded from books that popularized science as a vehicle of a materialistic *Weltanschauung*. Popular science loomed large in his eyes as the harbinger of blind determinism, intellectual imperialism if not plain tyranny, and of stifling hopelessness.

Chesterton learned about popular science not only by reading, as has already been mentioned, all Huxley and Spencer and everything else on the subject available in English, but also by moving about day after day in Fleet Street, the center of what we call today the media. The feeling of "nothing new under the sun" becomes inescapable as one reads his now fifty-year-old warning: "Let any one run his eye over any average newspaper or popular magazine, and note the number of positive assertions made in the name of popular science, without the least pretence of scientific proof, or even of any adequate scientific authority."[15] The avalanche of cures suggested for cancer and other ills in today's tabloids, together with the endless formulas offered for unusual mental powers, gives to Chesterton's lines a timely ring indeed. Science degraded to the level of quackery was not the only kind of popular science he took for his target. He found fault with an apparently better kind of it because it was not scientific, that is, logical enough, and he put the matter with thumping brevity: "I only pause . . . to show that even, in matters admittedly within its range, popular science goes a great deal too fast, and drops enormous links of logic."[16]

Rank disregard for logic in books on science written by men with proper scientific training could rankle him to an extent that may appear outright censorious. He called for "the dynamite of some great satirist like Swift and Dickens" to destroy "a certain kind of modern book which . . . ought to be blown to pieces." Not that he attributed bad faith to their authors. His immediate concern related to the fact that books contained "actually no scientific argument at all. They simply affirm all the notions that happen to be fashionable in loose intellectual clubs."[17] Yet, writing this around 1910 à propos the manifesto *Sexual Ethics,* by the famed Swiss neurologist Auguste Henri Forel,[18] Chesterton argued for much more than logic. He never ignored the existential havoc played by systematic disregard for logic which is the very essence of scientism. Today, when herpes claims its victims by the millions in the USA alone, where the number of families also has shown an alarming decrease for the past twenty years,[19] where the acquisition by homosexuals of full legal rights, including the right of shared property as in a marriage, encouraged pederasts to claim legitimacy—in these United States even some members of those "intellectual clubs" seem to be taken aback.[20] They no longer take lightly the disregard for logic which has been the mainstay of countless "scientific" books published during the last two decades on sexual behavior and related matters. Many of those outside such clubs are still to realize what is the lasting value of the kind of dynamite, which only a master of words like Chesterton can pack into his phrases and which alone is an effective weapon against a scientism that destroys man in the name of science.

Chesterton did not ignore interviews given by scientists on patently nonscientific topics, another form in which scientism demands disregard of the limitations of the scientific method. That demand, as Chesterton rightly saw, is dictated not by science but by a materialism which is really a pseudo-religion if not plain mystery-mongering.

Such was the gist of Chesterton's reaction to interviews in which Edison endorsed reincarnation in the name of science. In doing so Edison played a trick with the notion of the mind in the manner of a mystic and a mystagogue. The opening of headquarters for the Search for Extraterrestrial Intelligence (SETI)[21] could not have called for a better editorial than the rest of Chesterton's strictures of Edison, the materialist, parading in white lab coat: "He is a mystic because he deals entirely in mysteries, in things that our reason cannot picture; such as mindless order or objective matter merely becoming subjective mind. And he is a mystagogue because . . . he pontificates; he is pompous; he tries to bully or hypnotize, by the incantation of long and learned words. . . . "[22]

Sir Arthur Keith, a scientist of far greater stature than Edison, came in for Chesterton's strictures for much the same reason. This time the format within which logic was flouted in the name of science was not a book, not even an interview, but the far more glamorous lecture hall where one could play to the galleries and cover lack of logic with clever words. Keith became in Chesterton's eyes a preacher of popular science, who "throws a long word at us [and] thinks that we shall have to look it up in the dictionary and hopes we shall not study it seriously even in the encyclopaedia." Such was the tactic and manner of Keith's denial, in the name of science, of man's immortality, a purely philosophical matter. What made that tactic revealing for Chesterton, and this shows something of his interpretative power of the scientific scene, was its negative feedback on the presentation by such scientists of matters strictly scientific: "Even in those things he [Keith] betrayed a curious simplicity common among such official scientists. The truth is that they become steadily less scientific and more official. They develop that thin disguise that is the daily wear of politicians."[23]

The foregoing quotation may provide a natural transition to another major concern of Chesterton about science, namely, the pontificating, quasi-divine, and plainly despotic status accrued to it. From the vantage point of the 1980s, when the overweening role of science in practically all domains of life has become a commonplace, a prophetic ring may be detected in Chesterton's remark made around 1920. In that remark science, which "preaches the destiny without the divinity of Calvinism," is presented as one of the three chief institutions of the day. A prophetic ring may also be detected in Chesterton's singling out the two others: "A Divorce Court cutting up families with the speed of a sausage machine" and "a Finance that crosses all frontiers with the same enlightened indifference that is shown by cholera."[24] In far greater numbers than in 1920 could Chesterton today overhear two men talking, if not in a streetcar, in a faculty lounge or lab cafeteria: "Soon there would not be but the great Empires and confederations guided by the science, always the science."[25] Again, even more characteristic of our times would be his observation made in *Orthodoxy*: "Scientific phrases are used like scientific wheels and piston-rods to make swifter and smoother yet the path of the comfortable."[26]

The use of such phrases seems all the more unobjectionable, because science has taken on an undefinably vast existence of its own before which man tacitly bows: "Science, that nameless being, declared that the weakest must go to the wall; especially in Wall Street."[27] Thus "some men say that Science says this or that; when they only mean scientists, and do not know or care which scientists."[28] Particularly apalling in Chesterton's eyes was the ubiquitous invocation of Einstein's name. He saw in it a cultural curse to be resisted by the imposition of a small fine on anyone who after dropping Einstein's name "could not demonstrate before a committee of mathematicians and astronomers that he knew anything about Einstein." This dragging in of Einstein's name did not so much aim at voicing a considered judgement; rather it served mostly those who wanted to show off. The flood of books and symposia on Einstein's relativity, which three quarters of a century have seen steadily grow in number, contains plenty of illustrations for the accuracy of Chesterton's words: "Einstein is not part of any ordinary human argument, because any ordinary human being does not know where his argument leads or what it can really be used to prove. It may be, for all I know, a perfectly good argument for those who really follow it; but those who drag the name without the argument cannot know what the argument means."[29]

About these continual references to Einstein, Chesterton also made a remark that would have done credit to any perceptive historian of science. Contrary to a superficial look, the fashion of mentioning Darwin's name at a drop of a hat often implied some knowledge in biology. "Any number of people did really attack the study of biology, in order to agree or disagree with Darwin.... People did talk about Darwinism as well as about Darwin.... The talk about Darwin may have been popular science, but it was science, and it was popular." Quite different was the case concerning Einstein and relativity. "Hardly one person in a thousand thought of attacking the higher mathematics in order to agree with Einstein.... They know nothing but the name and the notion that something very important has happened in connection with the name.... The talk about Einstein may rather be called popular nescience." With his customary outspokenness Chesterton went for the jugular: "I believe that the reason is a certain increased laziness of the intellect; that fewer people are ready for a long sustained logical demonstration, quite apart from whether we think that the demonstration really demonstrates."[30]

It would have been, of course, too much to expect that Chesterton should perceive that objectivist and even absolutist core of Einstein's theory of relativity which began to be unveiled only from the late 1960s on by some perceptive historians and philosophers of science. That some of these are all too eager to take the really metaphysical sting out of their perception, while those who don't are given the silent treatment, would have certainly irked Chesterton. He repeatedly acknowledged Einstein's greatness as a scientist, but his common sense remained rightfully suspicious of the threat which customary phrasings of relativity posed to the truth of commonsense perception as the ultimate assurance of objectivity and normalcy. Logic itself was undermined when that truth was tampered with; the only safe antidote was, according to Chesterton, the reassertion by Christianity of the logic of understanding. For in that twentieth century, which "cannot as yet even manage to think itself anything but the age of uncommon nonsense," it is becoming increasingly clear that "in order to understand Einstein, it is necessary first to understand the use of understanding."[31]

Once the all-importance of commonsense knowledge is kept in mind, Chesterton's occasional barbs at Einstein and relativity will not appear the *lèse majesté* committed by a mere ignoramus or by a brazen upstart. It is well to recall that what Chesterton perceived with the unerring instinct of a born philosopher, however untrained, was emphasized by the professional philosopher, Bergson, who pointed out to Einstein, at the Sorbonne in 1922, that whatever the apparent overthrow of simultaneity by relativity theory, the latter presupposes it as its very foundation.[32] At any rate, the real target of those barbs was always something else. In one case it was rationalism wrapped around T. H. Huxley's advice to let reason carry us as far as it can go: Chesterton wrote in 1921 that "Science was supposed to bully us into being rationalists: but it is now supposed to be bullying us into being irrationalists. The science of Einstein might rather be called following our unreason as far as it will go, seeing whether the brain will crack under the conception that space is curved, or that parallel straight lines always meet."[33] In another case, the real target is the overly intellectualized character of old Greek Orthodox art, which for Chesterton "almost recalls the shameless mysticism of Professor Einstein, when he says, 'All space is slightly curved.' "[34] In another case, Einstein is brought in only because of the dubious status of literary parallels: "I have a deep and hearty hatred of literary parallels; specially when they have a suggestion of literary plagiarism. I object to the parallels on many grounds; but among others, on the ground that they are never parallel. Or, if we may (with all respectful allowance for Mr. Einstein) put the matter as a mathematical paradox, we might say that the two lines of thought are indeed parallel because they never manage to meet."[35]

While with respect to Einstein's relativity Chesterton, a layman, could have found even in its high-level popularizations mostly nonsensical conclusions, such as the abolition of the simultaneity of events and of anything non-relative, the situation was markedly different with respect to the great progress made in the 1920s and atomic physics. During the last ten years of Chesterton's life the English were treated to a series of brilliant science-popularizations by Eddington and Jeans, themselves first-rate physicists. An ever-recurring theme in those books was the untenability of materialist determinism as based on science. This could not fail to strike a chord with Chesterton. In his review of the six steps of his conversion, the fifth was devoted to "The Collapse of Materialism" as proclaimed by modern physical science. There is almost a ring of *nunc dimittis* in what he wrote in 1935, a year before his death, about a "world event," a "true revolution" in scientific thought. He wished it had come while he was still a youth, because it would have "hugely hastened" his conversion. He had obviously read Eddington, whom he found "more agnostic about the material world than Huxley ever was about the spiritual world." The remark revealed a judicious insight as to the limit set by realism, within which the new outlook of science could be a safe ally: "A very unfortunate moment at which to say that science deals directly with reality and objective truth."[36] He was certainly not to buy Eddington's idealism, if not plain solipsism, be it wrapped in science. Nor was he to miss his fun on observing old-guard materialists, who found it impossible that one could be a physicist without being a "scientist," that is, a spokesman for materialist determinism. Since the one who voiced that impossibility was Mencken, Chesterton could argue ad hominem: "I also am a scientist; but in my time it used to be called a journalist." At any rate, "the new physicists are not propagandists, but Mr Mencken so far from reverencing them as Science, desperately refuses to respect them even as scientists."[37]

The overweening tyranny of science, against which he chafed from his youthful years, now seemed to be overthrown: "The new scientists themselves do not ask us to accept the conclusions of science. The new scientists themselves do not accept the conclusions of the new science. To do them justice, they deny vigorously that science has concluded; or that it has, in that sense, any conclusions. The finest intellects among them repeat, again and again, that science is inconclusive."[38] He hoped that the lesson would not be soon lost on young scientists who came to the scene after the great change had been accomplished: "If the young scientist would ever allow us to regard his hypothesis as anything so human as a half truth, it might sometimes really be worthwhile to find the other half. If, instead of claiming that everything is covered by his explanation, he confined himself to pleading that there is something in his suggestion, he would look considerably less a fool when the next man, with the new explanation, comes along in about thirty years."[39]

Chesterton also hoped that Christians would not be so foolish as to base their faith on the new physics brought about by a scientific revolution whose bearing upon personal religion he found "often misstated and widely misunderstood." There was, he warned, nothing "particularly Christian about electrons," or for that matter "anything essentially atheistic about atoms." There was no use for slogans like 'electrons for the elect' and 'for priests and proton.' Christian philosophy was no more to be based on the new physics than ancient theology was to be propped up by recent biology. The new physics was of "catastrophic importance" for Catholics, only because it triggered a "collapse of materialism." Since by "catastrophic" Chesterton clearly meant "paramount," his reference to the possible collapse of "even the most confident cosmic statements of science"[40] will stand some clarification. He did not imply by that collapse that scientific statements contained no partial truth. Again, he was, as will be seen, too much a champion of the rationality and fundamental importance of the notion of cosmos as to mean by cosmic statement more than a sweeping claim or conclusion. Nor was he foreseeing an eventual exploding of the electron which would lead to no further knowledge. What he had in mind was a smug expectation that the last word had been spoken or would be spoken soon about the physical universe. Fifty years after Chesterton pointed his finger at that smugness, it is showing its perennial lure again in sanguine statements of many science writers about a Grand Unified Theory being around the corner, statements no doubt encouraged by some overconfident Nobel-laureate physicists who should know better.[41] Such smugness invariably betrays the presence of scientism.

At any rate, Chesterton's chief interest in science always centered on its possible threat to the freedom of the will. He perceived early in his career that scientific law as such was not a philosophical statement either on reality or on causality, but presupposed a realist philosophy if its statements were to have any bearing on either or both. Chesterton was rightly concerned about the broader pseudo-philosophical cultural context which turns science into scientism. He saw correctly that the revolution in physics thoroughly undercut the scientism that had set the tone of the days of his youth: "The Determinists of my youth used to boast that Science supported them, because some scientists talked about the Determinism of Matter. I do not know what they are saying now, when several scientists are actually talking about the Indeterminism of Matter." What he added in the next breath showed how firmly anchored he was in what really mattered: "Anyhow, the idea of choice is an absolute, and nobody can get behind it."[42]

Gone were the times when on account of their subscribing to deterministic philosophies Chesterton had to brand some men of science as being "almost barbarians."[43] He no longer had to contend with that scientific materialism which "binds the Creator Himself [and] chains up God as the Apocalypse chained the devil [and] leaves nothing free in the universe" as something worse only than that Calvinism which "took away the freedom from man, but left it to God."[44] While the climate of opinion could change, and at times perhaps for the better, to say as Chesterton did about scientific determinism that it was "simply the primal twilight of all mankind," was to express in an outburst a perennial truth.[45] Those outbursts of Chesterton will not even for a moment appear a hyperbolic abuse of rhetoric when set beside indictments that prominent modern physicists brought down on materialist determinism. One of them, Walter Heitler, saw in it the chief source of that demise of moral awareness that led to the two great wars of our century.[46] About both Chesterton predicted, it is well to recall, that they would make full use of the inconceivable horrors which technology can deliver and that any effort to keep them from enveloping the entire globe would be in vain.

In this century overawed and mesmerized by the marvels of science—in a sense they cannot be admired enough—nothing is more tempting than to make much of Chesterton's occasional barbs at science. His dictum that "physical science is like simple addition: it is either infallible or it is false,"[47] is a gross overstatement. His claim that "when science found that colours could be made out of coal-tar, she made her greatest and perhaps her only claim on the real respect of human soul,"[48] reflects the partiality of the artist he was. His aside that "art is long, but science is fleeting,"[49] may have been true for the past. Whatever the relative youth of science with respect to art, science is here to stay. Not that Chesterton expected science to disappear. The real aim of his barbs at science was the dislodging of scientism. Otherwise he would not have pleaded from the very start that "the rebuilding of this bridge between science and human nature is one of the greatest needs of mankind."[50]

Chesterton's celebration of literature over science should seem no more reprehensible than Matthew Arnold's animated defense of belles-lettres against T. H. Huxley's banishing Homer, Dante, and even Shakespeare from education on the ground that they do not prepare for life.[51] True, the study of literature does not provide the skill needed by master plumbers and computer programmers; it is still, apart from religion, the only repository of insights which alone provide meaning for technology and science. Chesterton had in fact an exalted notion of science and of great scientists. He contrasted Darwin's sincerity with the dogmatism of his disciples.[52] Even there he could make a generous exception, such as when he referred to Huxley as that "great and good man,"[53] and could rightly claim: "I have never said a word against eminent men of science."[54] As to science, he saw it an enterprise so noble as to be compared only to Christianity, or rather to the Christian creed, both in respect to structure and function. He saw in science a complex structure and praised scientists for being proud of it: "When one believes in a creed, one is proud of its complexity, as scientists are proud of the complexity of science. It shows how right it is in its discoveries. If it is right at all, it is a compliment to say that it is elaborately right. A stick might fit a hole or a stone a hollow by accident. But a key and a lock are both complex. And if a key fits a lock, you know it is the right key."[55] The reference to keys was not accidental. The complexity or high degree of specificity of Christian creed and ecclesial structure is nowhere more evident than in the keys entrusted to Peter, a point made by Chesterton on more than one occasion.[56]

This putting of science on the same pedestal with Christian creed will not appear a baroque rhetoric, if one recalls Whitehead's remark about the respective stirs made by the Babe in the manger and by the rise of science.[57] Chesterton saw in fact the functions of science and faith as analogous: "Science finds facts in nature, but Science is not Nature; because Science has co-ordinated ideas, interpretations and analyses; and can say of Nature what Nature cannot say for itself. The Faith finds its facts and problems in humanity, even in heathen humanity; because it brings to it principles of life and order and understanding, and comprehends humanity as humanity cannot comprehend itself."[58] And just as faith could work miracles, so could science. Unfortunately, science had no built-in safeguards to remain strictly scientific, a state which was Chesterton's professed ideal about science, one he wanted to safeguard with his barbs.[59] He had an uphill fight on hand. He had to recognize that "having worked its wonders, which are really on the material plane comparable to miracles, it [science] has gained a sort of glamour which is made to cover any number of trivial or disreputable conjuring tricks."[60]

Partly responsible for this were, according to Chesterton,

two factors, both inherent in the scientific development. One was the ever greater self-consciousness of science or rather of scientists. What Chesterton wrote in this respect deserves to be recalled at some length as it would do honor to the most perspicacious psychoanalyst of the scientific mind. In contrasting the great age of science, which he designated as the age of Darwin, with subsequent times, he noticed the latter grow weaker because of increased consciousness of strength: "Darwin was convincing because of his unconsciousness; one might almost say because of his dullness. This childlike and prosaic mind is beginning to wane in the world of science. Men of science are beginning to be aesthetic, like the rest of the world, beginning to talk of the creeds they imagine themselves to have destroyed, they are beginning to be soft about their own hardness. They are becoming conscious of their own strength—that is they are growing weaker."[61]

The other factor was the growing specialization in science. Chesterton put it bluntly: "Science means specialism and specialism means oligarchy." What this meant was the gradual decrease of the extent to which the body scientific could be checked from the outside. Long before it became fashionable to speak of the need of a democratic control of science and technology, Chesterton called attention not only to the immediate but also to the ultimate alternative: "The expert is more aristocratic than the aristocrat, because the aristocrat is only the man who lives well, while the expert is the man who knows better. But if we look at the progress of our scientific civilization, we see a gradual increase everywhere of the specialist over the popular function. Once men sang together round a table in chorus; now one man sings alone, for the absurd reason that he can sing better. If scientific civilization goes on (which is improbable) only one man will laugh because he can laugh better than the rest."[62]

Only by ignoring this context can Chesterton be pictured as an unbeliever in the future of science. Of course, he was never the kind of believer in science who finds not only a scientifically triggered Armageddon impossible, but also, who, like Jacob Bronowski and other pundits, absolves science and scientists of any and all responsibility for Hiroshima.[63] They in fact do not even want to acknowledge far lesser evils that science can produce when turned into a gullible handmaid of scientism. The outcome was branded by Chesterton as a cultural curse. According to him the process "could be expressed with terrible exactitude in one phrase: Science has become a name to conjure with."[64] This was the science that Chesterton rendered in *Eugenics and Other Evils* as "stuffy science," as "modern craze for scientific officialdom," as "bullying bureaucracy and terrorism of tenth-rate professors."[65] Their resolve to reshape society by a so-called "scientific sociology" prompted him to speak of the "great scissors of science." The immediate context of that priceless phrase was Chesterton's letting his wrath descend on the "sociological doctors" who in the name of public hygiene proposed that schoolgirls coming from poor homes with no bathrooms have their long braids, their sole pride, cut short.[66] Were Chesterton alive today, he would perhaps speak of the great "spark plugs of science" which drag small schoolchildren on three-to-four hour round-trip bus-rides daily in the name of cultural hygiene. And upon learning that the some of the same pundits responsible for that legislation came to discover that the educational chances are largely decided by the quality of fairy tales a child is told when not yet three, Chesterton would now see fully vindicated his extolling literature over science, to say nothing of scientism.

The way he stood up for literature could be startling. In his first novel, *The Ball and the Cross*, he put the point across by siding with the humanist MacIan and not with Turnbull, the physicist, as the two argued whether the views of one or the other were more effective in filling mental asylums to capacity.[67] As it turned out, the hatred fanned by men of letters would not have been overly effective in turning first Europe, then much of the world, into a vast morgue, had scientists not provided new means of mass murder. They did so with apparently small revulsion, if Oppenheimer's famous plea of "not guilty" can be taken as typical. He justified the construction of the bomb not only as a wartime necessity but also as a "technically sweet" project with which scientists would go ahead regardless of other considerations.[68] Such "technically sweet" projects are today offered by genetic engineering, which may trigger greater havoc than atomic bombs. The apparent inability of society to exercise control over technology evokes that madness which Captain Ahab confessed to in a phrase that would have done credit to Chesterton: "All my means are sane, my goal and object are mad."[69] Chesterton himself said much the same when he denounced in *Heretics* "the task so obviously ultimately hopeless, of

49

using science to promote morality."[70] In speaking so, Chesterton was not any less scientific than say an Einstein, who a dozen or so years after him warned that not the uranium but the human heart needed to be purified, that he himself could not derive a single ethical value from science, and that scientists as such were not a whit more imbued with ethics than were other men.[71] Objection to all this comes not from science but from scientism.

If man was to be saved, it was of crucial importance to keep returning, as Chesterton did, to the theme that science had its limits, the very point which scientism tries to make one overlook and wholly ignore. This point should seem of no less significance for science itself if there is as much as a grain of truth in the warning of no less a physicist than James Clerk Maxwell: "One of the severest tests of a scientific mind is to discern the limits of the legitimate application of the scientific method."[72] By assigning unlimited relevance and competence to the scientific method, scientism rules out precisely that test. By setting quantitative exactitude as the only and supreme test of truth, scientism robs of meaning the world of qualities and values. By the same stroke it makes meaningless science as well. The process was aptly described by Chesterton: "Science which means exactitude, has become the mother of inexactitude."[73] For when the instruments, however precise, of science are put to work on a patently qualitative target, they will appear crude and blunt. Such targets ranged, according to Chesterton, from man's wish for a pork-chop to man's desire for heaven! All such longings and their objectives had in them something intrinsically indefinable and variable, which turned them into unsuitable objects for exact science: "This kind of vagueness in the primary phenomena of the study is an absolutely final blow to anything in the nature of science. Men can construct a science with very few instruments. . . . A man might measure heaven and earth with a reed, but not with a growing reed."[74]

There is more depth in that single statement than in all books written by positivists on the philosophy of science. Perhaps some philosophers of science took note, through Gardner's book, of Chesterton's masterful unmasking of the exalted notion of scientific law. In all likelihood they remained unaware of the foregoing passage from *Heretics*, which might have produced in their humbler kind the same sense of puniness which Gilson, a prince of Thomists, felt on reading Chesterton's *St. Thomas Aquinas*. Their logical positivist kind would certainly have recognized in that passage a devastating indictment of their supreme tenet, an endorsement of scientism, that everything has to remain on the surface, to recall the motto of the Vienna Circle.[75] It is to the professional superficiality, where exclusive attention is given to quantities, that scientism wants to reduce man so that he may be turned into a one-dimensional being.[76]

Chesterton battled not science but scientism because he wanted to save man. The essence of his strategy was to claim for science and its products that genuine admiration which scientism could not provide: "The wrong is not that engines are too much admired, but that they are not admired enough. The sin is not that engines are mechanical, but that men are mechanical."[77] Such phrases, which put in sparkling relief man's non-mechanical abilities and therefore his uniqueness, were so many saving graces for many a young man looking for an escape from the hold of scientism. One of them, who spent the years 1909–12 at Harrow and Balliol, recalled: "Mr. Chesterton's destructive criticism of the Huxleys, Bradlaughs, and Haeckels of our youth was as devastating as it was brilliant, and its value would be more widely appreciated today if it had not been so completely effective."[78] Unfortunately, scientism is a hydra that can forever grow new and more numerous heads. It should be enough to think of the actual wide use of euphemisms which wrap even the worst sins in "Latinate verbiage for true obfuscation,"[79] so that the veneer of scientific respectability may cover festering boils. There is an eery timelessness in Chesterton's exposing the trick, which turns habitual theft "into hunger for monotonous acquisition" and polygamy "into an enhanced development for the instinct for variety." His biting comment, that "science has views broader and more human," clearly was not out of place.[80]

Cyrus Pym, the scientist in *Manalive* who viewed "things in abstract," was no caricature. He still embodies all those scientists who preach, like Monod, that all is chance or necessity or both.[81] Cyrus Pym is still a perfect stand-in for those scientists who, like I. I. Rabi, exalt science, which alone has the imagination to reveal in things the glory of God, and debunk arts and letters, including Shakespeare, as "mere gossip which does not take us outside ourselves."[82] Cyrus Pym still foreshadows such a highly publicized pundit of scientism as Jacob Bronowski, who put up a spirited defense on behalf of the uniqueness of man only to conclude that man is uniquely animal.[83] Cyrus Pym is still the paragon of those scientists who, while disclaiming any part in the "scientific" dehumanization of man, are all too eager to use a terminology which in itself is the denial of any depth, responsibility, and life. And also of romance, Chesterton's favorite term, when speaking of the fullness of life: "The romance of conscience has been dried up into the science of ethics. . . . The cry to the dim gods, cut off from ethics and cosmology, has become mere Psychical Research. Everything has been sundered from everything else, and everything has grown cold. . . . This world is all one wild divorce court." Such a graphic snapshot of scientism had to come from the one who more effectively than anyone battled it with all the literary and philosophical powers at his disposal. He was an implacable antagonist of those who in the name of science—be it anthropology, sociology, or psychology—applied their dissecting scalpel to the wholeness of man, and above all to the wholeness of man's mind. He fought them because his own mind was once in the grip of their destructiveness. He also fought them because, as will be seen, he had grippingly vivid insights to tell about the mind's unique powers. He fought because he had hope. He felt that "there are many who still hear in their souls the thunder of the authority of human habit: those whom Man hath joined let no man sunder."[84]

CHAPTER THREE

Critic of Evolutionism

Inserting a part of "The Ethics of Elfland" among great essays on science took some daring, as its author "was not noted for his knowledge of things scientific." In saying this Martin Gardner was certainly accurate. What he said in the next breath, "Chesterton never, for example, could quite bring himself to accept the theory of man's descent from lower animals," was wholly inaccurate for more than one reason.[1] If Gardner had brought up, say, relativity, quantum mechanics, or electromagnetic theory for an example, he would have accurately illustrated his statement insofar as it suggested unfamiliarity on Chesterton's part with things scientific, given that Chesterton enjoyed little or no reputation for expertise in such things. Actually, the example, as phrased by Gardner, implied a good deal of familiarity on Chesterton's part with evolutionary theory, or with Darwin's theory of man's descent, to be specific. Moreover, it was most inaccurate to suggest, as Gardner did, that those who could not quite bring themselves to accept that theory served evidence of their ignorance of it, as if knowledge of that theory imposed its acceptance on any well-informed and logical mind. There were, in Darwin's time and have been ever since, first-rate biologists, fully familiar with that theory, who found great faults with it. Still others did not accept it at all without thereby becoming anti-evolutionists.[2] At any rate, whatever the extent of the disagreement of those experts (of whom more will be said later) with Darwin's theory, that is, his very specific notion of the sole mechanism of evolution, Chesterton never did as much as come close to accepting it. In fact, he roundly rejected it, precisely because he formed for himself a most considered judgement on that miscegenation of very good science and very bad philosophy which is Darwinism. As such a union Darwinism would deserve to be seen as the intellectual counterpart of the pairing of a fine horse with a lame donkey, were it not for the fact that unlike their offspring, the notoriously barren mule, Darwinism propagates itself with unslackening vigor.

The best known and most restrained form of Chesterton's judgement on Darwinism is in his *Everlasting Man*, one of his great masterworks if not the greatest. Sections of it should long ago have entered anthologies on anthropology and evolutionary theory. Compilers of such books shy away from Chesterton, though he is not a completely unknown entity to them. In fact, in a recent and prestigious textbook on prehistoric anthropology, the chapter on the origins of culture is headed by a Chestertonian parody of the much vaunted transition, also called missing link, from animals to man: "Man is an exception, whatever else he is. If it is not true that a divine being fell, then we can only say that one of the animals went entirely off its head."[3]

Chesterton would not have been Chesterton if he had not come up time and again with other, even more biting remarks about a creature which had a built-in fascination on account of its chronic elusiveness. The best of those remarks is from those pages of *The Everlasting Man* where he unburdens himself on what he found most burdensome about Darwinism and all: the illogical dogmatism of Darwinians. Reluctant to follow their master's sincerity in acknowledging the missing status of the Link, they "have insensibly fallen into turning this entirely negative term into a positive image. They talk of searching for the habits and habitat of the Missing Link; as if one were to talk of being on friendly terms with the gap in a narrative or the hole in an argument, of taking a walk with a nonsequitur or dining with an undistributed middle."[4]

Nothing would do more injustice to Chesterton than to present him an enemy of evolution insofar as it merely claims many transitional forms and therefore a very long geological past. Had he denied either or both claims, his name would be today on the lips of creationists. Whatever the staunch Roman Catholicism of Chesterton, he would not on that account be an overly compromising ally for creationists and Fundamentalists, who are not embarrassed to call the witness of such debunkers of metaphysics (and of scientific progress as well) as Sir Karl Popper and Thomas S. Kuhn.[5] Fundamentalists invoke these two, because their theories of science are logically destructive of even that hallowed scientific theory, Darwinian evolution, which they hold to be the only explanatory framework appropriate for any branch of thought.[6] Fundamentalists, or creationists, who bring to light with the diligence of an army of ants any testimony, however questionable, on their behalf, know, I presume, that Chesterton could happily live with two things that are both anathema to them: biological gradations, however countless and imperceptible, and pre-historic times, however far they carry us into the geological past. Whatever Chesterton's unabashed reference to God, they can be but dismayed with remarks of his such as, "If evolution simply means that a positive thing called an ape turned very slowly into a positive thing called a man, it is stingless for the most orthodox; for a personal God might just as well do things slowly as quickly, especially if, like the Christian God, he were outside time,"[7] or "Evolutionists cannot drive us, because of the nameless gradation in Nature, to deny the personality of God, for a personal God might as well work by gradations as in any other way."[8] Fundamentalists would find these remarks a sword that cuts both ways. This is why they would not find much solace in Chesterton's protest against "the custom to make fun of Fundamentalism and to suggest that [fundamentalist] American religion is rather antiquated." In the rest of the protest Fundamentalists appear slightly less unsavory than their chief antagonists, the secular humanists, to use a neologism well postdating Chesterton, who merely said: "American irreligion is much more antiquated than [fun-damentalist] American religion and the sceptic can be more of a fossil than the sectarian." Chesterton, who never skirted an issue, let alone the fundamental issues, noted about Fundamentalists not only that "they are funny enough," but that "the funniest thing about them is their name. For whatever else the Fundamentalist is, he is not fundamental. He is content with the bare letter of Scripture—the translation of a translation—without venturing to ask for its original authority."[9]

Just as in theological matters Chesterton's thinking was always in the grip of ultimate issues, his dicta on evolution were riveted in the fundamental points of fact and logic. It should therefore be of no surprise that the most basic formulation of his methodology in interpreting science, and the science of evolution in particular, is contained in his *Everlasting Man,* his most fundamental theological work. The methodology imposed on him, on the one hand, the acknowledgment that having no claim to learning he would have to depend for his data on those who are more learned, that is, the specialists. On the other hand, he would claim "the reasonable right of the amateur to do what he can with the facts which the specialists provide."[10]

The word amateur would be the least significant part of that statement—a sort of transparent tactic to secure the aura of professionalism—had Chesterton not made a reference in the next breath to H. G. Wells's *Outline of History.* He described it as the work of another amateur relying for his facts on the specialist and claimed to himself the right to do the same. But there was much more to that reference than meets the eye. *The Everlasting Man* left the printing press toward the end of 1925. Early that year there appeared the third "revised and much more amply illustrated" edition of Wells's *Outline.* Two years earlier, in January 1923, it had seen its second edition, "revised very severely and rearranged," to quote again

Wells's own words.[11] By then Wells's book had taken the public by storm and found among its many reviewers Chesterton,[12] who was coming closer and closer to his step of entering the Catholic Church. The step was anything but a small quiet move at the end of a long deliberate march. While it was not an agonizing step rankled by doubts, overwhelming it certainly was to the point of shaking Chesterton to the very core of his being.[13] Those who noticed more astringency and less joviality in the older Chesterton, and blame this shift on his conversion, seem to have no eyes for its depth.[14] They would be well advised to look for a prophet who kept joviality for his major key.

The Everlasting Man was Chesterton's intellectual reaction both to the experience of his conversion and to the stir created by Wells's *Outline of History*, a book, it is well to recall, written when Europe lay prostrate as never before and optimism based on science lost for a while its luster. In trying to understand that colossal debacle and collapse, Wells was led farther and farther back in history, first to the Enlightenment and then to the Middle Ages. He soon perceived that not even the Roman Empire would give him a self-explaining starting point. Rome called for Greece, which in turn evoked Persia and older civilizations. Wells was soon speculating on prehistoric man and beyond him on the evolution of mammals and all their predecessors. Finally, he stopped at the formation of the solar system which, unknown to him, had just turned for scientists into a most uncertain topic. All that would have hardly spurred Chesterton into replying not only with a book, but with a book which remained his most sustained systematic argumentation on any topic. The spur was Wells's section on Christ, a section much shorter than the one which dealt with, say, the struggle of the Greeks with the Persians. Clearly, such disparity was glaring if Wells really meant that Christ was the outstanding teacher of mankind.[15] Not only did Christ not stand out in the rest of the *Outline;* He was not visible at all. As to Christianity, it was invisible once Wells left the Middle Ages behind, which he did very quickly.

Had Wells come out with his *Outline* ten years earlier, or around 1910, *The Everlasting Man* might not have been written at all. The Chesterton of 1910 and for some time later was the Chesterton of an "orthodoxy" curiously generic. The philosophy pervading his *Orthodoxy* was, of course, the kind of realist metaphysics on which alone can orthodox dogmatics be safely erected. But some very orthodox dogmas, among them the divinity of Christ, were hardly visible in Chesterton's *Orthodoxy.* On the contrary, his *Everlasting Man* is an outline of history of which Christ, the God made Man, is the very rationale and culmination. So much for the background history of Chesterton's writing the *Everlasting Man;* the significance of that background for Chesterton's dicta on Darwinian evolution will be taken up later.

For the moment let us return to Chesterton's methodology which, as was noted, had philosophical realism for its backbone. He *really* wanted to *do* something with the facts and only with *real* facts, that is, facts obvious, easily accessible, towering facts. Doing something *real* with *real* facts imposed on him the strict avoidance of mere verbalism.

He was to shy away from big words as if their bigness were an explanation. Evolution was such a big word. Not that he went as far as George Bernard Shaw, whom he held high as the Saint George that had slain the polysyllable to the point of discussing evolution without mentioning it.[16] But Chesterton at least made very clear at the outset that the word evolution became the vehicle of illusion whenever it stood for the belief that by attributing extreme slowness to a process one answered the question why things were proceeding, let alone progressing, in the sense of evolving. Taken for that belief the word evolution was for Chesterton a skirting around "the ultimate question" of "why they [things] go at all; and anybody who really understands that question will know that it always has been and always will be a religious question; or at any rate a philosophical or metaphysical question."[17]

Chesterton was not, of course, so naïve as to think that a pointing, however forceful, at the fundamental character of a thing, that is, at its involving a realist metaphysics, would impress most scientists. Scientists blandly presuppose the existence of things and limit themselves to the quantitative and descriptive aspect of things, which is least apt to activate the mind's readiness to ponder reality as such, that is, to do good metaphysics. Chesterton therefore quickly passed on to the facts, and as a good strategist he centered on a single striking fact. One breakthrough, if it is good enough, is more than enough to route the opposition.

The opposition, or the so-called "integral Darwinism" or integral evolutionism, would not have become the secular creed if it somehow had not been turned into a mimicry of the Creed, which is a profession of the Triune God. Integral Darwinism has indeed from Haeckel's time displayed a cunningly pseudo-Trinitarian character. Its three divinities are the inchoateness of the universe, the spontaneous origin of life, and the automatic rise of human consciousness. To the credit of Darwin, he did not discourse on cosmic origins, although it is not difficult to guess what he would have said. He would have repeated Herbert Spencer, whom he took for one of the greatest philosophers of all times,[18] a judgement which tells as much about Darwin's philosophical discernment as about Herbert Spencer's philosophy, cosmic or other. As to the origin of life, Darwin felt that, mysterious as it could appear, it had to be wholly natural.[19] Concerning the origin of human consciousness and all that goes with it, *The Descent of Man* was in Darwin's eye *the* reply, and an unabashedly materialistic reply at that.[20] One wonders what his reaction would have been, had he lived to hear Sir Andrew Huxley, President of the Royal Society and certainly a Darwinist with the best family credentials, warn in late 1981 his Darwinist colleagues against taking lightly the problem of the origin of life and the origin of consciousness.[21] Such a warning amply reveals the persistent absence of a scientific solution to these pivotal questions. The witness of the latest in modern scientific cosmology about the emergence of the universe will be taken up in the next chapter. So much in a way of background to Chesterton's strategy to be offered by a historian of science interested in the philosophical convictions of scientists, evolutionists or others.

Without indulging in the pleasure of sketching the evolutionary mimicry of the Triune God, Chesterton clearly identified the three cardinal dogmas, if not deities, of evolutionism. He must have meant only philosophers of the more perceptive type when he stated in *The Everlasting Man* that "no philosopher denies that a mystery still attaches to the two great transitions: the origin of the universe itself and the origin of the principle of life itself. Most philosophers have the enlightenment to add that a third mystery attaches to the origin of man himself. In other words, a third bridge was built across a third abyss of the unthinkable when there came into the world what we call reason and what we call will." And he added: "Man is not merely an evolution but rather a revolution."[22] The word revolution calls for comment. The political history of our lifetime made revolutions an almost daily affair. Worse, a revolution took place also in the academia in the sense that, owing to the cleverness of some philosophers and historians of science basking in the tranquillity of the academy, revolutions have become the principal explanatory device that need not be explained, not even properly defined. In fact, the count of intellectual revolutions is now so large that the few periods left for normalcy may appear as revolutions. Chesterton was a man of his word also in the sense that when he used a word he took it in its pristine and forceful meaning. He meant to be taken for a true revolutionary when he said that the emergence of man was a Revolution.

He felt he could prove this with only one fact. As it happened not infrequently, he was not exact in details. He suggested that the cave serving as background in the chapter

"The Man in the Cave," of *Everlasting Man,* was first entered by a boy and a priest. Such was not the case with the caves of Altamira, discovered in 1878, nor, as it seems, with the caves of Font de Gaume, near Les Eyzies, first explored around 1903, which secured genuineness to the paintings in Altamira. But the facts inside the caves of Font de Gaume, of which a priest, Henri Breuil, was among the first to inform the world in a massive study,[23] were too gigantic to be reduced by Chesterton's poetic inaccuracies. After all, he himself added to those magnificent paintings an interpretative portrait, no less original and no less gigantic in its terseness and cultural significance than the paintings themselves. Once more he said in a few words what in the hands of others would lead to long paragraphs and even longer chapters. What the boy and the priest saw "were drawings or paintings of animals and they were drawn or painted not only by a man but by an artist. Under whatever archaic limitations they showed that love of the long sweeping or the long wavering line, which any man who has ever drawn or tried to draw will recognize and about which no artist will allow himself to be contradicted by any scientist."[24] Chesterton certainly was not afraid to stand up to certain scientists; they were to be resisted.

What followed was a route of a materialistic notion of man on the basis that no animal except man made symbols and lived by them. Half a century after Chesterton declared, "art is the signature of man,"[25] not a single counterevidence has turned up to weaken, however slightly, that classic definition along either of two parameters emphasized by Chesterton. First, he stressed the uniqueness of artistic creativity which, in his inimitable diction, established man as the only creature who is also a creator.[26] In this emphasis many noted philosophers such as Alfred North Whitehead, Susanne Langer, and others followed him, not, of course, to the extent of stating unequivocally the uniqueness of man or to the extent of giving credit to Chesterton. Not at all a surprise to that Chesterton who once noted about a great Catholic discoverer that he was "to be Catholic first and to be forgotten."[27] From coaxing birds to paint and gorillas to talk, every laboratory trick of experimental psychology has been tried to turn up a counterevidence but in vain. No evidence turned up to discredit either the other parameter specified by Chesterton, the suddenness of the onset of creativity as displayed by paleolithic paintings. This is not to suggest that Chesterton made much of their age, fifteen or twenty thousand years. All he stated was that the evidence indicated a suddenness: "Monkeys did not begin pictures and men finish them. Pithecanthropus did not draw a reindeer badly and Homo Sapiens draw it well ... the wild horse was not an Impressionist and the race-horse a Post-impressionist ... there is not a shadow of evidence that this thing evolved at all. There is not a particle of proof that this transition came slowly, or even that it came naturally. . . . In a strictly scientific sense, we simply know nothing whatever about how it grew, or whether it grew or what it is. . . . It was not and it was; we know not in what instant or in what infinity of years."[28]

Were he still alive, Chesterton would chuckle on reading in a recent book on the subject that the prehistoric period, now largely synonymous with Lascaux, emerges in fact as the empirically attested period of that suddenness. Even more so would Chesterton find significant some further facts about that book. Its author, who never made a secret of his crusading materialism, had wished nothing more than to find and present a counterevidence. In its absence he was forced to celebrate that suddenness as the *Creative Explosion,* the very title of his book, an inquiry into the origins of art and religion, that is, into the origin of the very humanness of man.[29] Clearly, such a witness is very valuable but also very true to himself. Chesterton lived before the times when the terms creation and creativity had become a hobbyhorse and lost their basic meaning through constant wear and tear. The academic penalty for that loss is being paid in some prestigious universities in the form of remedial courses in English for those who are enrolled in graduate courses in creative writing. Had Chesterton known these times, he would have come up with a memorable verbal blast against those who brandish the weapon of creative explosion and carefully muffle its sound in the same move. Perhaps, instead of a blast he would come up with a roar of laughter. For it was with a reference to man's ability to laugh that he brought to a close "The Man in the Cave," the opening chapter of *Everlasting Man:* "Alone among the animals he is shaken with the beautiful madness of laughter; as if he had caught sight of some secret in the very shape of the universe hidden from the universe itself."[30] And as an afterthought he referred to man's ability to feel shame, which he defined as "the presence of some higher possibility," a classic in the forcefulness of understatement.

In seeing the almost systematic slighting of such obvious human characteristics as laughter and shame, a slighting practiced by almost all protagonists of man's strictly animal origins, one does not know whether to laugh or to hang one's head in shame. For it is both hilarious and shameful (the compound of these two produces the explosive tragedy) that so obvious characteristics of man, which encompass his very wholeness, are being ignored in favor of far less significant and far less obvious details. This drastic disproportion evidenced in Chesterton's eyes a well-nigh blindness on the part of evolutionists: "The evolutionist stands staring in the painted cavern at the things that are too large to be seen and too simple to be understood. He tries to deduce all sorts of other indirect and doubtful things from the details of the pictures, because he cannot see the primary significance of the whole. . . . To see man as he is, it is necessary . . . to keep close to that simplicity that can clear itself of accumulated clouds of sophistry."[31]

Indulging in sophistry was in Chesterton's eyes one of the lesser evils of evolutionism, though very evil in the long run. He found an example of sophistry in the injudicious use of the term pre-historic. In the case of man it should have implied the realization that pre-historic was largely synonymous with unknown, precisely because it referred to time antedating recorded history. Bafflingly, the painstaking patience of scientists was not in evidence when it came to pre-historic times and man, which turned Chesterton's wit mordant: "Sometimes the professor with his bone becomes almost as dangerous as a dog with his bone."[32] This vignette, if not immortal at least incomparably graphic, and now almost sixty years old, would easily prove its timelessness through disclosures about the haste with which the discoverers of "Lucy" rushed to their conclusions.[33] If not the commercial, at least the intellectual pattern is now secular, that is, stretching beyond a century. It took a Chesterton to diagnose it in a few lines which shall not be improved upon: "Doubt and caution are the last things commonly encouraged by the loose evolutionism of current culture; . . . the one thing that it cannot endure is the agony of agnosticism. It was in the Darwinian age that the word first became known and the thing first became impossible."[34]

With that we may safely leave behind *The Everlasting Man,* because it was not there that Chesterton excoriated another major sophistry of evolutionary parlance, namely, natural selection. It implies, together with "struggle for existence" and "survival of the fittest," some sort of an intelligence embedded in a Nature written with a capital. In a brief remark Chesterton labeled natural selection as "an extravagantly improbable conjecture,"[35] of which more later. In a longer comment he described it as a chief example of the perpetual falling back by scientists on sheer mythology. Chesterton was not, of course, the first to point out that personification of Nature or of any of her powers had no room in scientific theory, let alone in Darwinism, and that the survival of the fittest was mere tautology. But once more he said all this with inimitable directness: "Nature selecting those that vary in the most successful direction means nothing whatever except that the successful succeed" and "the whole Darwinian argument is that it is not a case of Nature selecting, any more than of God selecting, or any one else selecting, but a case of things falling out in that fashion."[36] For all its hollowness this mythologizing of nature keeps itself alive in Darwinian discourse. It is a cheating which can take such proportions as to alert even an enthralled television audience, such as the one which in both hemispheres watched week after week "The Life on Earth" series conducted by David Attenborough. He received numerous protests for regularly falling back on the phrases "it developed" or "they developed," taken transitively, as he presented with extraordinary skill the even more extraordinary devices observable in countless species, devices which would put to shame our most skilled solid state engineers, aerodynamicists, and radar specialists.[37] It should seem indeed puzzling that after more than a hundred years of Darwinism no substitute has been found to a parlance that in itself proves there is purposeful action in at least one species, Homo sapiens, a sheer incomprehensibility within Darwinism.

A third sophistry that Chesterton found a convenient target was the claim of most supporters of Darwinism that it is a theory with no serious problems. That in the 1920s Darwinism caused plenty of headaches to not a few of its prominent students could not remain hidden even to amateurs. To say, as Chesterton did around 1930, that "all men of science have abandoned Darwinism," let alone that "all men of science have abandoned Materialism,"[38] was a hyperbola, though not as hollow as it may appear. Precisely because of the revolution in physics, which Chesterton enthusiastically welcomed, only a few scientists retained allegiance to the mechanistic materialism of Haeckel and others. As to Darwinism proper, the Chestertonian hyperbola should seem a flat phrase when compared with the verdict of professors of the Natural History Museum of Paris, who declared in 1935, in the fifth volume of the *Encyclopédie Française:* "It follows from this presentation that the theory of evolution is impossible. . . . Evolution is a kind of dogma in which its priests no longer believe but which they keep presenting to their people."[39] To be sure, Darwinism was not dead to the extent to which this was, around 1922, suggested by Chesterton in an essay with the title, "Is Darwin Dead?"[40] Nor was natural selection as much without evidence as Chesterton claimed there, although then as now much could be made of the bafflingly small evidence for transitional forms. Were he alive today, Chesterton certainly would not be surprised to find some young Darwinists who, in order to cope with the absence of those very gradual transitional forms, simply abolish the need for them. He would tell them that in speaking of punctuated evolution, that is, of an evolution which works not slowly but in repeated sudden bursts so as to run quickly through many transitional forms, they offer mere verbalism. Its purpose is to draw attention away from very serious problems as if those problems did not exist.

Today, four generations after Darwin, prominent biologists are not lacking who candidly admit the grave difficulties of Darwin's theory. There was more than one such prominent biologist in the 1920s and 1930s.[41] Indeed, the theory of descent by natural selection is unique among major scientific theories. No other theory has produced for now over four generations a persistent and scientifically most respectable minority of dissenters, who, as one of them put it, refuse to become "Lamarckian fellow travelers."[42] They are Darwinians who, together with their credulous counterparts, were given by Chesterton a marvelously accurate portrayal: "As a fact, modern evolutionists, even when they are still Darwinians, do not pretend that the theory explains all varieties and adaptations. Those who know are rather rescuing Darwin at the expense of Darwinism. But it is those who do not know who doubt or deny; it is typical that their myth is actually called the Missing Link."[43] The majority speaks as if all going were smooth. In the 1920s one of its representatives was Sir Arthur Keith. Chesterton's raking him over hot coals is worth recalling for two reasons, both symptomatic of a situation not at all limited to the 1920s. One reason has to do with Chesterton's unmasking in an essay, "The Mask of the Agnostic," the motivation which, as Chesterton put it, "puts up Sir Arthur Keith to deny that there is any change in the scientific attitude about Darwin."[44] The motivation he saw was Keith's rank materialism, which made him declare before the British Association that the jury of biologists, swearing by facts alone, disproved the possibility of survival after death.[45]

It remains to be seen whether there will be enough historians of science in these days of ours, when the field all too often turns into a psychoanalytic exercise, who would have the courage to probe into the conditioning by a scientist's materialism of his scientific theories. (Darwin was not the first major modern student of natural history who almost from the start subscribed to rank materialism.)[46] It will take no less courage on the part of such historians to face up to the rejection of their manuscripts by leading publishers and periodicals, if they were to unmask another pattern which relates to the second reason for which Chesterton's reaction should seem instructive. That in denying the change of attitude toward Darwinism Keith ignored recent and prominent evidence was bad enough; worse, his gaffe was not to be reported in the organs of the "establishment" he belonged to. The fact and evidence found publicity only in the columns of a Catholic weekly. Unfortunately, too many so-called Catholic weeklies or monthlies would be found by Keith today to be tacit allies of what Chesterton called "party papers" of a science harnessed into the service of materialism: "Probably the story [of Keith having been proved sensationally and disastrously wrong] is now known to all readers of that paper; but it will possibly never come to the knowledge of most other journalists, and it certainly will not be recorded in most of the other papers; and they support the party leader when he publishes the official contradiction. They will not let the public know how triumphantly his other contradiction was contradicted."[47]

The most self-destructive sophistry in descent by natural selection *alone* related, and still does, to what it can state in terms of its logic about the transformation of one species into another (to say nothing of the fearfully mysterious transitions among orders, classes, and phyla)[48] the very thing which Darwin meant by the "origin of species" whose existence he simply took for granted. The problem as late as the early 1970s was so serious that a very learned though patently Darwinian survey of the logical problem of species and its transformation ended with the suggestion that "such a thing as a metaphysics of evolution as a *biological phenomenon* is not only desirable but also necessary."[49] The author of that survey also blamed Darwin's antagonism to anything savoring of metaphysics for the chronic neglect by biologists of what is implied metaphysically when they speak of species. While Chesterton did not discuss the species problem as such, he did not miss at all its very basis. The latter consists in what Chesterton summed up in 1925 in a priceless phrase about the "grey gradations of twilight" which "Darwinists everywhere suggest because they believe it is the twilight of gods."[50] Long before that he was fully aware of the Darwinist sophistry which implies the systematic elimination of sharp contours, distinctness, and definitions while holding high the eternal chaotic flux of all into all. Herein lies the real crux that the notion of species represents for that very Darwinism which tried to explain it by explaining it away.

One of Chesterton's early essays deals with this problem, and does so in fact not without a metaphysical, nay theological, hint at the species problem. The year is 1905 and the place is Edinburgh's central feature, the Scott Monument. With its extraordinary richness of contours carved into stone it did not fail to strike a chord with Chesterton, always fascinated with the strictly circumscribed and definite. As he looked at those sharp contours and exact details, he saw in the distance the heights of Arthur's Seat. Unlike from a distance, Arthur's Seat was, at a close range, but "vague curves of clay, vague masses of grass." Only the monument embodied the kind of precision that conveyed "certainty, or conviction, or dogma, which is the thing that belongs to man only, and which, if you take away from him, will not leave him even a man." Such was the bias on which Chesterton declared it to be "the whole business of humanity in this world to deny evolution; . . . the business of the divine human reason to deny that evolutionary appearance whereby all species melt into each other." From philosophy Chesterton quickly went to theology in the form of a powerful evocation: "This is probably what was meant by Adam naming the animals."[51] The transformation of a tree into a cloud with no center, no edges, and no tangible features whatever, was another parable of Chesterton's to conjure up the abolition of identity in which he saw the chief threat of evolutionism to sanity.[52] A shorter and more direct statement of his on the same threat is a part of his description of the fantastic flying ship of Professor Lucifer in *The Ball and the Cross:* "Every sort of tool or apparatus had . . . to the full, that fantastic and distorted look which belongs to the miracles of science. For the world of science and evolution is far more nameless and elusive and like a dream than the world of poetry and religion; since in the latter images and ideas remain themselves eternally, while it is the whole idea of evolution that identities melt into each other as they do in a nightmare."[53]

It would be most tempting at this point to reply to all this with a stentorian "enough!" and rise in a solemn protest against this vilification by Chesterton of the sacred science of evolution. But before we throw the whole book at him, especially all the books of Father Teilhard de Chardin, Chesterton must be given a fair hearing. Indeed, he has much more to say and even more deprecating things about evolutionism, which, it is most important to recall in order to assure him a fair trial, he does not deplore à propos the long geological past or the countless transitional forms (provided they are found in a number even vaguely proportional to what they have to be if Darwin is right). Chesterton deplores evolutionism because it abolishes forms and all that goes with them, including that deepest kind of ontological form which is the immortal human soul. He deprecates evolutionism also because it dilutes and saps the existential defenses of the individual. Therein lie the roots of what Chesterton identified as the "subconscious popular instinct against Darwinism." It was not the "offense at the grotesque notion of visiting one's grandfather in a cage in the Regent Park." All normal men could for a practical joke make beasts of themselves as well as their grandfathers. The real reason was rather their realization that "when once one begins to think of man as a shifting and alterable thing, it is always easy for the strong and crafty to twist him into new shapes for all kinds of unnatural purposes."[54]

All the specious efforts to erect a respectable ethical system on Darwinism were unmasked as Chesterton remarked: "Darwinism can be used to back up two mad moralities, but it cannot be used to back up a single sane one. . . . On the evolutionary basis you may be inhumane, or you may be absurdly humane, but you cannot be human. . . . In neither case does evolution tell you how to treat a tiger reasonably, that is, to admire his stripes while avoiding his claws."[55] Nothing indeed can be more effective in sapping moral energies and in making moral decisions impossible than an outlook in which everything merges into everything. In considering the merit of injecting evolutionary thought into every part of the educational program Chesterton noted prophetically that "it would not make that education very insistent on the ideas of free will and fighting morality; of dramatic choice and challenge."[56]

The systematic or rather inescapable blurring of sharp contours, physical as well as moral, within the evolutionary perspective implied the abolition of any Mr. Jones: "being within the scope of evolution . . . his edges are rubbed away." Significantly, Chesterton stated this as an afterthought to his observation that evolution as such posed no threat to the existence of God, "who can work by gradations as in any other way."[57] The specific thrust of Darwinism aimed at man before aiming at God: "Evolution does not specially deny the existence of God; what it does deny is the existence of man."[58] Concerning the threat of "stronger races" against the "weaker" races, a threat sanctioned by Darwinism (though talked away by most Darwinists uneasy about Darwinism being seen for what it is), Chesterton's vision was largely limited to the programs of eugenics through control of heredity and scientifically arranged marriages. Such programs were proposed in large numbers and with small acumen in the England of Galton and his followers. Chesterton had more than one *bon mot* for them. Heredity, in his view, was "trumpeted aloud by men who did not even know what it meant, long before even the most learned men knew even the little about it that they know now. Mendel the monk has explained it much more fully than did Darwin and all the material scientists; but even after that it remains a very obscure and subtle subject, as the scientists would be the first to admit."[59] Very timely words in this age of genetic engineering. As to scientifically controlled marriages, it was in his eyes a theory that "could only be imposed on unthinkable slaves and cowards." To be sure, he did not fail to notice that some very selfish motivation was lurking in the background whenever the theory of races was put forward: "Of all forms in which science, or pseudo-science, has come to the rescue of the rich and stupid, there is none so singular as the singular invention of the theory of races."[60]

Chesterton was too much of a patriot and too much of a Francophile to perceive the Anglo-Saxon origin of the message of the "charlatan Haeckel" on whom German militarists fell back for ideological support and who, to Chesterton's consternation, had a large following in his merry England. To spell out the harsh truth was left for George Bernard Shaw, who in the wake of World War I gave the fearfully accurate diagnosis and prophecy: "We taught Prussia this religion; and Prussia bettered our instruction so effectively that we presently found ourselves confronted with the necessity of destroying Prussia to prevent Prussia destroying us. And that has just ended in each destroying the other to an extent doubtfully reparable in our time."[61] This somber truth may not sound too far-fetched in these very years which kept little if any of the euphoria of the 1960s when the ushering of a golden age seemed to be around the corner. Gone are the years when the naïve optimism of Teilhardian evolutionism gained hosts of no less naïve converts. When the going is really rough and the prospects patently bleak, there is less willingness to repose on the crest of the great evolutionary wave, biological and social. Hence, there may be a realistically perceived and experienced ground to appreciate Chesterton's strident reaction to evolutionism, instead of brushing it aside as the old-guard Catholic's miscomprehension not only of what is new in science but also of what is already fairly old in it. To be sure, nothing would be more mistaken and unfair than to downplay the role played by Chesterton's Catholicism (both incipient and full grown) in his reaction to Darwinian evolution where his reaction to scientism also comes to a head. Unfair it certainly would be to that Catholicism which is not the brainchild of Johnny-come-lately theologians but the well-proven challenger of already two millennia. Many centuries before integral Darwinism turned man, body and soul, into a chance product of purely accidental physical forces, true anthropology was defined in the dogma of Incarnation.

To downplay that Catholicism would certainly be unfair to Chesterton, whose *Everlasting Man* culminates in soaring chapters on Christ, the God become Man. Whatever the strictly demonstrative value of those chapters, they could only be written by someone who was in the grip of Christ's superhuman reality taken in the dogmatically most orthodox sense. That sense stands or falls with certain tenets about Christ's human nature which, as any other such tenets, have their inner logic. One of those tenets is that Christ had an immortal *human* soul. It is upon that soul, inseparably united to his divine nature and divine person that Scripture and the Creeds predicate Christ's descent into hell. Such a point will sound quite Chestertonian as soon as one recalls a remark of Chesterton in *The Ball and the Cross*. There the question of Turnbull, the scientistic scientist, "what is the difference between Christ and Satan?," is given the reply: "It is quite simple. Christ descended into hell. Satan fell into it."[62] Chesterton knew all too well the difference of Christ and the difference He makes.

A chief difference is that Christ's very being contradicts any doubt about man's being radically different from the rest of creation and even from the rest of the evolutionary process. Christians have sensed this difference long before Darwin, for whom, let us not forget, the expectant look of a dog at its master was not essentially inferior to the gaze of a man worshipping on his knees.[63] The same difference is still sensed by Christians who, long after Darwin and in spite of Father Teilhard's vagueness on the subject, profess to have an immortal soul. For them it is not a mere logical *non sequitur* but a blasphemy to hear the claim that the only thing that differentiates man from animals is the far shorter time which it takes for man to turn to his own advantage the messages channeled by his feedback mechanisms. According to that claim, at least in its most outspoken form for which one ought to be grateful to Professor Philip Morrison of MIT, termites, if given time enough, would come up with telescopes.[64]

Now, in this age of anti-intellectualist fideism nothing is more fashionable than to look at the Christian's acceptance of Christ as a mere matter of faith. Chesterton's whole discussion of Christ in *The Everlasting Man* meant to show that the acceptance of Christ can be, and indeed has to be, a rational recognition of His uniqueness as Man, before He as God becomes the object of faith. Again, in Chesterton's argument, the empirical uniqueness of Christ, the Man, is the empirically most effective safeguard, declaration, and reminder of the uniqueness of man. *The Everlasting Man* provides the insights which alone can do justice to the tension arising from the fact that individuals and races are, for all their equality, very different. Whenever that equality, which has to be everlasting if it is to be true equality, is denied to man, as is the case with Darwinism, ominous consequences loom large. They can at best be talked away by well-meaning Darwinists with a lame reaffirmation of the essential equality of all races and individuals.[65] Consistent advocates of Darwinism have no choice but to applaud, as Darwin did,[66] the crushing of "lower" nations by "higher" nations.

Chesterton had a lynx's eyes for spotting most hallowed tenets which modern man would have had to repudiate had he obeyed the logic of his admiration for Darwinism. One such tenet, that all men are created equal, was expressly stated in the American Declaration of Independence and received its most memorable encomium in Lincoln's Gettysburg Address. In bringing to a close his reflections on his first American tour with an essay on the future of democracy, Chesterton called attention to the contrast between Lincoln's "invoking a primitive first principle of the age of innocence" and the tacit expectations of his nation's intellectuals, ready to advance from pragmatism to social Darwinism: " 'We hold these truths to be probable enough for pragmatists; that all things looking like men were evolved somehow, being endowed by heredity and environment with no equal rights, but very unequal wrongs', and so on." The moral imposed itself with ironclad rigor: "I do not believe," Chesterton added, "that [such a] creed, left to itself, would ever have founded a state; and I am pretty certain that, left to itself, it would never have overthrown a slave state. What it did do . . . was to produce some very wonderful literary and artistic flights of sceptical imagination. . . . All had grown dizzy with degree and relativity. . . . There were different sorts of apes; but there was no doubt that we were the superior sort."[67] A more ironic and accurate portrayal of the concrete logic of Darwinism will probably be never forthcoming.

Logic had to be the chief arguable point about Darwin and evolutionism for anyone so profoundly attached, as Chesterton was, to Christ, the Logos. Typically, his essay "Is Darwin Dead?" comes to a close with the remark: "I am very far indeed from calling the Darwinian a liar; but I shall continue to say that he is not always a logician."[68] It was on that ground that he kept pointing at the geological record as the Achilles' heel of an evolution based on chance mutations and natural selection alone. That heel is still in need of a thorough immunization against a large number of hazards if the very recent debate on one of the greatest questions of evolution, the origin of flight, has any message at all. The debate still finds evolutionists divided into two camps according to whether flying originated from trees down or from the ground up.[69] Both camps are, of course, at one in postulating countless small steps and transitional species to bridge that gap in either way. They are also at one in rejecting one another's flights of fancy for imagining those transitional mechanisms. Most important, both camps are at one in having only meager evidence on hand for transitional forms. Since 1861, when in a quarry in Bavaria workmen found on a split limestone what became called *Archaeopteryx*, the much touted (and much disputed) transition between reptiles and birds, few more such specimens have turned up in spite of intense search. The situation is so frustrating as to percolate down to the level of science reporting. On reading one such report on "the little solid evidence to fill in the great blanks of avian evolution,"[70] Chesterton would feel himself entitled to repeat: "If the proofs of natural selection are lost, why then, there are no proofs of natural selection; and there is an end of it."[71]

I shall end this chapter with a warning against taking Chesterton as a guide concerning the technicalities of evolutionary science or of any science for that matter. He would have been the last to claim such a role for himself. While he would smile at seeing the substantive lack of fossil evidence ruffle the feathers of students of avian evolution, he would not be ruffled to see the evidence now available on behalf of some very limited evolution for which natural selection may be a sufficient explanation.[72] With his penetrating logic he would point out that this new development, largely postdating his death, has not made a dent on the immense disproportionality between the amount of proofs desired and the proofs available. His mind was always riveted on keeping such and similar disproportionalities in focus. His hyperbolic dicta make sense only if they are read with that sense of proportion which alone makes sense.

Nowhere is that precept of greater importance than in a context of achieving a sensible view of man and evolution. For if it is true, as T. H. Huxley once admitted, that seeing continuous gradations among all species, classes, orders, and phyla is a metaphysical vision,[73] then the question rightly arises about the proportion between what is empirically seen and what is intellectually envisioned. Had Darwinians remembered Huxley's remark and, what is far more important, had they done what Huxley himself failed to do, namely, to ponder what is a metaphysical vision, Chesterton or others would have had no real cause against evolutionary science. But if it is true that man is able to form a metaphysical vision, then nothing is more important than to ponder that factor in man which makes him able to see beyond his own physique and anything physical. In fact, hyperbolic statements alone can drive home the immense disproportionality between what man sees here and now, and what he can perceive beyond things, nay that everything which is the universe.

Man's ability to see the Beyond is man's saving grace, worth any effort to keep it vigorous. Chesterton may have done mere apologetics by gearing all his dicta on science to the cause of saving man, but he did not have to, nor did he ever apologize for having done so. He was an insightful student of evolution and a pitiless critic of evolutionism, because his belief that man was worth saving was more than a humanistic cliché. He did not have to apologize for seeing clearly that unless man was secure, his worshiping of the Creator was most insecure. And since logic, if true, must be universal, unless man was secure, there could be no secure scientists, not even Darwinists. Finally, man's security was also the only safeguard of speaking safely about the universe, this most sublime and most fundamental object for science. As will be seen, the universe had few finer champions than Gilbert Keith Chesterton.

CHAPTER FOUR

Champion of the Universe

As Chesterton arrived at the threshold of his adulthood, the precocious sensitivity of the genius he was made him stare across another, far more portentous line, the one that separates being from non-being, or rather, what is much the same, the vast realm of things from the bailiwick of the mere self. In speaking of Chesterton or of any topic dear to him or concerning him, the topic of solipsism presents itself almost inevitably. The depth of his youthful fathoming of the abyss, which is solipsism,[1] can be gathered from the spontaneity with which he kept returning to it for the rest of his life. He saw in it the real and ever threatening alternative to normal existence. For him solipsism was never an abstract theoretical issue, but an issue most existential. His most forceful statements on it are either in his novels or in his literary criticism, as it was for him a truism that often reality is most concretely seized through literature or reflections on it. Among the latter was his essay on "The Orthodoxy of Hamlet" dating from 1907. For Chesterton, Hamlet's orthodoxy and his greatness as a thinker consisted in not being a skeptic. He was great enough "to know that he was not the world. He knew that there was a truth beyond himself....The real sceptic ... sinks through floor after floor of a bottomless universe."[2] Twenty years later this immediate connection between reality and the universe was also prominent in *The Poet and the Lunatics,* a book aptly described as containing Chesterton's best appraisal of solipsism.[3] There Chesterton makes Gabriel Gale remark: "I also dreamed that I had dreamed of the whole creation. I had been behind and at the beginning of all things; and without me no thing was made that was made. Anybody who has been in that centre of the cosmos knows that it is to be in hell."[4]

In reflecting on this and similar passages nothing would be easier than to overlook the invariable presence there of the terms universe and cosmos, which may not after all be very logical. The obvious contrast to solipsism is belief in external reality and not necessarily an assertion of a universe. But with a Chesterton, for whom words were not playthings but basic tools for getting a hold on reality, habitual use of the word universe could not be a mere habit. In fact, for him the universe was the only strict alternative to solipsism and to its logical corollary, pessimism. In the very process in which during his early adulthood he shook off pessimism, he joyfully embraced not merely optimism but that very reality which is the universe. Embracing was once more an enthusiastic acceptance of something given. A letter of his to his most trusted friend, Edmund C. Bentley, written when he was about twenty, puts the whole matter in a wording which may strike the fastidious as a mixed metaphor, whereas it conveys a most

88

extraordinary intimation of the metaphysical contingency of the universe: "A cosmos one day being rebuked by a pessimist replied, 'How can you who revile me consent to speak by my machinery? Permit me to reduce you to nothingness and then we will discuss the matter.' Moral. You should not look a gift universe in the mouth." Similar gems are contained in the Notebook, the record of his first literary gropings. There the invitation sent out by Gilbert Chesterton "requesting the pleasure of humanity's company at tea on Dec. 25th, 1896," comes to a close with a list which includes not only Humanity Esq. as one of those invited but also "The Earth" and last but not least "Cosmos E." In "Prayer of Man Resting," also in the Notebook, the word of thanks addressed to the Lord is preceded by the line: "My head is bowed before the Universe." Maisie Ward also found noteworthy another passage in the Notebook in which young G.K.C. called the attention of a careless multitude to "that great Empire upon which the sun never sets. I allude to the Universe."[5]

Such a youthful attentiveness to the universe, the most encompassing object available for man, necessarily propelled Chesterton toward developing a philosophy in which the real universe played a pivotal part. The development was so fast as to have had an irresistible force behind it. The earliest writings of Chesterton abound in startling statements on the universe as primary object for philosophical reflections about existence itself, and on the high stake the universe represents in the great ideological contest.[6] Parties to that contest were not only the scientistic materialists but also the pantheists. It was the challenge from a pantheist, a certain Mr. Rix, that first brought out of Chesterton the most profound aspect of his dicta on the cosmos. The challenge, which related to Chesterton's claim that everything was divine, was taken up by him in an essay, "The Temple of Everything" (published on March 24, 1903, in the London *Daily News*), of which only a fourth, and hardly the best fourth, has so far been put again into print, after a few lines from it were quoted with an emphasis on their importance.[7] On reading the full essay one cannot help being seized with a mixture of astonishment and frustration. The frustration relates to the realization that hundreds of Chesterton's essays in the *Daily News* are still to appear in book form. Even if only one out of ten of those essays is so profound and brilliant as the "Temple of Everything," the lack of easy access to them should be considered a cultural deprivation all the more unforgivable, as any and all the stray dicta of far lesser authors have already found their way into genuinely complete editions of their works.

The astonishment engendered by this essay has to do with Chesterton's finding fault with pantheism not so much because it presents a false God, but rather because it logically must render man insensitive to the world and make human perception so nonsensical as to deprive man of his senses to the point of driving him mad. Chesterton, who always delights in unfolding in a picturesque manner the absurd position of his antagonists, does the same here too with the pantheist who can hardly remain sane if he is to worship, say, his collar button, and all the time at that. No less facetiously, Chesterton reminds the pantheist that "the attempt to prostrate oneself to one's own backbone is attended with pain and an indescribable sense of failure." But the real failure of pantheism implied much more than the limits of one's own contortional powers, physical or mental. The real failure was a self-defeating limitation of any attempt that wanted a limitless grasp of all: "When people say all things, they do not mean all things in the least. The Universe may include every detail according to logic; but it does not do so according to the human imagination. Let some excellent Pantheist . . . think in continents . . . let him summarise like a column of figures the incredible myriads of ages. And then let him lay his hand upon his heart, and tell me privately whether his vision really included a mental picture, let us say, of Lord Goschen's aunt. . . . And people who think of the All, and only of the All . . . are worshipping something heartless, brainless, bodyless, something that is everything and nothing, something that has not the power of giving anyone that shock of reality which we can get from a woman's face or a sting of pain."

In other words, Chesterton found fault with pantheism because it made one insensitive to that evidence of the senses which always presents us with particulars. These, if stared in the face, appear queer in the sense of defying any and all expectation and are therefore "nonsensical." This is the consideration that lurks between the lines in a little remembered passage from a still earlier essay of Chesterton, "A Defence of Nonsense," which fortunately found its way into the *The Defendant*, his first collection of essays from the *Daily News*. In that essay there is a phrase that anticipates the gist of Chesterton's best dicta on the universe: "If therefore nonsense is really to be the literature of the future, it must have its own version of the Cosmos to offer; the world must not only be tragic, romantic, and religious, it must be nonsensical also."[8] Later, in a more formally philosophical vein, he would celebrate the strange, almost nonsensical particularness of every being, including the universe, a particularity that calls for nothing less as an explanation than a true Creator, who happened to choose one specific cosmos from among an infinitely large number of possible universes.

In the foregoing passage there are two key parts for an understanding of the genesis of Chesterton the cosmologist. One part is, of course, the nonsensicality or utter particularity of the cosmos and of everything in it; the other is the concern for the literature of the future. The concern was that of one who for the previous three years or so had been earning his living as a journalist and literary critic. Criticism always implies a standard or ideal, however implicit. Chesterton, about to break into the literary scene, not only possessed that ideal but was thoroughly possessed by it. The proof of this is the promptness with which he took up Robert Blatchford's challenge to reveal his own creed after having picked to pieces many others' creeds.[9] The result was a rewriting of a dozen or so critical evaluations of literary figures, all called heretics, who had one heresy in common: their subjective estheticism. As Lynette Hunter aptly put it, Chesterton's charge against all of them—and they were a motley group: Shaw, Wells, McCabe, Kipling, Yeats, and others—was that the "esthete's belief that emotion originates in [the] self leads him to think that he personally can control it completely. He recognizes no external value in actual things because he invests them completely with his own."[10] Chesterton's own summary of their common heresy was more specific and also more extensive: "They are seeking under every shape and form a world where there are no limitations—that is, a world where there are no outlines; that is, a world where there are no shapes. There is nothing baser than *that infinity*. They say they wish to be as strong as the universe, but they really wish the whole universe as weak as themselves" (italics added).[11]

Clearly, if the real alternative to solipsism was the universe, the latter, taken in all its robustness, had to have as its alternative the debilitating weakness of solipsism. A page earlier, it is well to recall, there appears what seems to be Chesterton's first reference to Thomas Aquinas. We are in 1905. Chesterton has just turned thirty and also turned down an invitation to a chair of English literature at the University of Manchester. He was to serve the cause of the universe, not of a mere university. Those who are apt to take this for a cheap alliteration had better turn to the December 12, 1903, issue of the *Daily News*. There in the first heat of his reaction to the challenge of Blatchford, a well-meaning atheist,[12] Chesterton made his stand on nothing less than the universe, a subject immensely wider than the very vastness of English literature: "If Christianity should happen to be true—that is, if its God is the real God of the universe—then defending it may mean talking about anything and everything"—which, of course, Chesterton could do less unhindered as a journalist than as a professor of English literature, let alone as an academic eunuch parroting forever the actually prevailing consensus.[13]

The chief witness about the pivotal role that the universe played in Chesterton's philosophy or message from the very start is the introductory chapter of *Heretics,* which unlike the other chapters was not a reworking of an essay already published. In writing that chapter Chesterton clearly felt that he must give to the whole work its *idée maîtresse.* It was an orthodoxy that could be the counterpart of all heresies that agreed not to agree, and had for their only rule that there was no rule, not even a golden rule. They all were caught in a universal inconsistency because, as Chesterton put it, "they do not agree in their theory of the universe." Their common tenet—"cosmic truth is so unimportant that it cannot matter what any one says"—was a far cry even from the stance of old Liberals, let alone from the belief of those, among whom Chesterton counted himself, "who think that the most practical and important thing about a man is still his view of the universe. . . . We think the question is not whether the theory of the cosmos affects matters, but whether, in the long run, anything else affects them." If such was the case, all the standard-bearers of orthodoxy could logically expect that their ultimate task would be a crusade on behalf of the universe. So it was stated in the phrase concluding the *Heretics,* a phrase to which a consideration of the latest in scientific cosmology will force us to return: "We shall be left defending, not only the incredible virtues and sanities of human life, but something more incredible still, this huge impossible universe which stares us in the face."[14]

If the universe was the chief bone of contention or, if one prefers, the great continental divide between heretics and orthodoxy, then in a Chestertonian discussion of orthodoxy the universe could be expected to loom large. Indeed it did. The presentation of the universe as a cosmic town both queer, that is, so specific as to border on queerness, and very homey, that is, something very familiar and inviting on mere sight, was the objective of *Orthodoxy* set by its author. Once more he recalled solipsism as the alternative where, in the absence of a cosmos, thought itself was reduced to nullity: "If the cosmos is unreal, there is nothing to think about."[15] As to the choice between pessimism and optimism, a vote for the latter was not a matter of subjective taste but of a loyalty, the commitment of a cosmic patriot.[16] Herein lies the explanation of a most unusual sequence in that chapter, "The Ethics of Elfland," where a cosmology follows Chesterton's famed appraisal of scientific laws. There was nothing new, except the vibrant vividness, in Chesterton's stripping scientific laws of all bearing on reality. Positivists—Comte, Mill, Pearson, Poincaré, and Duhem—had done it already. Of these only Duhem insisted on the realism of common sense as the sole means through which scientific laws can be connected to the real world, but even a Duhem could not shake off all skepticism about the universe as such.[17] Chesterton was alone in moving directly and without hesitation from a strictly positivist preamble not only to a reassertion of reality, but to an assertion that embraced the entire cosmos. The second and larger half of "The Ethics of Elfland" is a cosmology, the chief anathema to positivists, crude and refined, traditional and logical. Logical positivists, had they flourished during the prime of Chesterton's life, would have certainly been singled out by him as the most heretical of all heretics and, above all, as their most illogical brand. As to those positivists who see their chief glory in falsifications, he would no doubt brand them as the most subtly destructive among all falsifiers. Indeed, there is hardly anything so falsely in place in modern philosophy of science as Popper's declaring, as if it had been his invention, the old truth that all science is cosmology.[18] For whatever the vastness of science as a conceptual system that can be accommodated in falsificationism, the latter secures not a grain of reality, let alone a real cosmos. Even in his latest on the subject, a book called *Open Universe*, Popper is very shy on the universe precisely because his aim is to make his reader believe that the universe is open to limitless possibilities and will be the realization of all of them.[19]

Such is, of course, the grand old illusion of a mere logician, who can speculate on castles rotating on a rooster's leg and other dreams that logic as such cannot find fault with. That they dream also of other universes, though this is a contradiction in terms, tells even more of their lack of logic. The realist Chesterton had no license for such mental extravaganzas. While for scientists, who merely take reality for granted, and for logicians, who play a mere game with reality, the universe is but a theory,[20] for Chesterton the realist it was truly a thing which, like all other things, was strictly limited both in itself and in its possibilities. Just before he turned to "The Ethics of Elfland," he made a statement which contains more truth than all books written by idealist philosophers: "The moment you step into the world of facts, you step into a world of limits."[21] This statement which followed another no less incisive one, "Art is limitation, the essence of every picture is the frame," was worthy of the future interpreter of Thomas and of the *form* that turns the "substratum" into a *substance*, that is, a concrete being.[22] Form is all the more appropriately expressed as frame, because if a framed picture is dropped, almost invariably it is the frame that breaks. The frame or form, being so specific, is fragile as all reality is. No wonder that in that cosmological part of the "Ethics of Elfland" Chesterton, in recalling the fairy tale about a glass castle, recalled also that reaction of his as a child which anticipated the best cosomological insight: "When the heavens were compared to the terrible crystal, I can remember a shudder; I was afraid that God would drop the cosmos with a crash."[23] Fragility was also the chief point as he recounted a few years earlier the story of the glass walking stick. If there was a queer specific thing here it was, pulled out from a dusty attic as a contribution to a benefit-sale to help the poor of the parish. It was all the more specific because it was filled from end to end with colorful pieces of candy. It is such an object for which one readily reaches out and which one readily breaks. For Chesterton it was the perfect symbol of the universe.[24] For all he knew, he saw and heard of only one such walking stick, just as he knew of only one fragile universe. The more fragile it was, the more precious it looked: "The universe is a single jewel and while it is a natural cant to talk of a jewel as peerless and priceless, of this jewel of the cosmos it is literally true. This cosmos is indeed without peer and price; for there cannot be another one."[25]

This may be too much of the metaphysics of cosmic contingency, but let us not forget that the very existence of a science that wants to retain empirical basis is based on the contingency of any and all. Otherwise there would be no need of laboratories and observatories but only of a priori minds, which are, of course, far less expensive commodities to have on hand. Startling as it may appear, there have been in this century quite a few first-rate astronomer-cosmologists who professed themselves to be solipsists or at least idealists in the Kantian sense. They were never so consistently idealist as to miss an opportunity to clamor for better and bigger telescopes, although they seized on any opportunity to reduce stars to a mere sensation on their retinae.[26] Chesterton not only was immune to such philosophical bungling, but he was also the first-rate philosophical cosmologist who instinctively made the proper improvement right there where some of the best philosophers went wrong in cosmology. I doubt that Chesterton ever read Book Lambda of *Metaphysics,* where the pantheist Aristotle roundly declared that the universe is a house without a master, or an army without a commander.[27] Chesterton's universe explicity had a captain and a "divine captain" at that, and this is why it had a Flag.[28]

All this, so it may seem, brings in natural theology, but where is science or rather the science of cosmology to be specific? Indeed, nothing is more relevant here than the question, Where is scientific cosmology? When Chesterton wrote *Orthodoxy,* cosmology as science was non-existent, insofar as cosmology means a scientific discourse about the universe, that is, the totality of consistently interacting things. There had, of course, been long before Chesterton scientists and astronomers who wrote on the universe, but what they offered was not scientific to the elementary degree of being free of contradictions or paradoxes that should have been obvious. The notion of an infinite homogeneous universe, which came into vogue only in the second half of the nineteenth century and which was never proposed by Newton, implied a grave paradox: the gravitational potential was infinite at any point of it. Such a universe cannot exist.[29] Partly under the impact of this consideration it became fashionable toward the turn of the nineteenth century to divide the universe into two parts, one observable, the other forever beyond observation. That the observable part was equated with the Milky Way should reveal how revisable can be a scientific consensus.[30] More important, this division of the universe into two parts was a perfect case of a schizophrenia which turns its victim's gaze away from an obvious difficulty in the hope that it will then go away. Such is hardly the attitude compatible with science.

As to speculations about the origin of the solar system, of which a good number had been proposed in the decades preceding the writing of *Orthodoxy*, they were wrongly called cosmogonies because they related only to puny parts of the cosmos and not to the cosmos itself.[31] When in a few cases it was attempted to extend the model of the genesis of the solar system, usually called nebular hypothesis, to the entire universe, the typical starting point was an infinitely extended gaseous nebula. Knowledge of it was completely nebulous. Its only distinctive feature was its wholly nondescript character. The presumed beginning of the infinite universe was the kind of homogeneity which is the very opposite of specificity. If Chesterton had anything scientific in mind, it was *that* kind of infinite homogeneity when he contemptuously referred to *"that infinity"* (italics added).[32] It deserved the most emphatic intellectual contempt, and the same holds true of those unphilosophical efforts in which our most specific universe is derived from utter non-specificity. The most noted, if not notorious, among them was the cosmogony of Herbert Spencer who, in H. G. Wells's blunt portrayal, "believed that individuality (heterogeneity) was and is an evolutionary product from an original homogeneity, begotten by folding and multiplying and dividing and twisting it, and still fundamentally *it.*"[33]

As I noted in chapter 1, Chesterton was not ignorant of the main developments of modern physics. In the England of Eddington and Jeans it was not difficult to learn about the skepticism that developed among physicists concerning materialistic determinism. He registered in the early 1930s, without being overawed, that "it was Eddington . . . who used the phrase that the universe seems to be more like a great thought than a great machine."[34] Chesterton needed no science to acquire that genuine philosophical vision which makes one discern lucid intelligibility even in dark matter. Nor did his rational assurance in the createdness of the universe need the news about the expansion of the universe. In fact, that assurance of his was so firm and profound as to enable him to put prophetically on the spot latter-day mystery-mongering in scientific cosmology, such as the inflationary universe, where the automatic turning of nothing into a cosmos is being celebrated: "For those who really think, there is always something really unthinkable about the whole evolutionary cosmos, as they [the Evolutionists] conceive it; because it is something coming out of nothing; an ever-increasing flood of water pouring out of an empty jug.... In a word, the world does not explain itself, and cannot do so merely by continuing to expand itself. But anyhow, it is absurd for the Evolutionist to complain that it is unthinkable for an admittedly unthinkable God to make everything out of nothing, and then pretend that it is *more* thinkable that nothing should turn itself into everything."[35] For another and related mystery-mongering a fine antidote can be found in Chesterton's words: "With all the multiplicity of knowledge there is one thing happily that no man knows: whether the world is old or young."[36] This dictum, which antedates by a quarter of a century the discovery of the expansion of the universe, contains more lasting truth than all the learned superficialities in which the absolute age of the universe is confidently estimated, as if there could be a physical state beyond which scientists could only spot the mere nothing.

While the two foregoing statements of Chesterton are isolated, this is not the case with his preference to consider the universe as a small thing, or rather to find the universe in very small things. One reason for this was his love for reality. Love always treats its objects as if they were delicately small. In *Orthodoxy* Chesterton spoke in this sense of the universe: "I was frightfully fond of the universe and I wanted to address it by a diminutive." In viewing everything as small, the artist only imitated God: "Even those dim and shapeless monsters of notion which I have not been able to describe, much less defend, stepped quietly into their places like colossal caryatids of the creed. The fancy that the cosmos was not vast and void but small and cozy, had a fulfilled significance now; for anything that is a work of art must be small in the sight of the artist; to God the stars might be only small and dear like diamonds."[37] A year earlier he had already described that artistic bent in terms of a contrast with the mathematical mind: "The artist sees things as they are in a picture, some nearer and larger, some smaller and farther away: while to the mathematical mind everything, every inch in a million, every fact in a cosmos must be of equal value. That is why mathematicians go mad, and poets scarcely ever do. A man may have as wide a view of life as he likes, the wider the better; a distant view, a bird's eye view, but still a view and not a map. The one thing he cannot attempt in his versions of the universe is to draw things to scale."[38] There has been, of course, more artistry in exact science in general and in scientific cosmology in particular than Chesterton, an outsider, could surmise, although science, for all the artistic ingredient in its creativity, remains very different from art. Still, in view of what happened in scientific cosmology, as will be seen shortly, during the last twenty-five years, there may be some relevance for science in what Chesterton said about artistic imagination: "The true spiral of imagination and creation is always twisting inwards toward smaller and smaller things, ever since men realized that jewels were smaller than pebbles and seeds smaller than clods; that if there is indeed a progress of humanity it may be such a progress to discover its own heart."[39]

While there can be no question about Chesterton's preference for the small, it should seem doubtful that his delight in seeing the universe in the diminutive and reduced to the scale of the small should be taken for a Thomist trait.[40] If such indeed were a Thomist trait, one could reasonably expect to find a trace of it in Chesterton's *St. Thomas Aquinas*. To be sure, the cosmos, which comes up there only once, is spoken of in a form that deserves reflection: "The nineteenth century left everything in chaos; and the importance of Thomism to the twentieth century is that it may give us back a cosmos. We can give here only the rudest sketch of how Aquinas, like the Agnostics, beginning in the cosmic cellars, yet climbed to the cosmic towers." The cosmic climbing of Aquinas was summed up in an immortal outburst of Chestertonian profundity: "There is an Is!"[41] It is subsequent to Chesterton's declaration that long before the child knows that grass is grass or self is self, he knows that something is something. He knows the *ens* or being. "Upon this sharp pin-point of reality," Chesterton continues, "Thomas rears by long logical processes, that have never really been successfully overthrown, the whole cosmic system of Christendom."[42]

A pin-point base never has been wider and more encompassing. As Thomas, Chesterton too, saw the universal in the singular. For both, by far the most universal and for the human mind the most universal entity had to be *being* itself. Since the universal could be grasped only through the singular, that is, in the specific, the totality of being, or the universe, had also to be specific and coherently so. This meant a specific order, the very opposite of chaos. The kind of chaos Chesterton had in mind may be guessed from a remark of his in *The Man Who Was Thursday:* "The rare strange thing is to hit the mark; the gross obvious thing is to miss it. Chaos is dull; because in chaos a train might go anywhere—to Baker Street or Baghdad. But man is a magician and his whole magic is in this that he does say 'Victoria' and lo! it is Victoria."[43] As to the nineteenth-century version of chaos, possibly he meant politico-social theories, all of which had their bases either in Hegelianism or in positivism or in materialism, certainly chaotic bases. If he had in mind nineteenth-century cosmological theories too, especially the one proposed by Herbert Spencer, he would have been certainly in his right to put the label chaotic on those theories. The assumption that he may have thought of such cosmologies or cosmogonies has some basis in a remark he made in 1903 in an introduction to an abridged edition of Boswell's *Life of Johnson.* There he spoke of the practical necessity of presenting an author through selections from his works, and noted that even such an interesting object as the earth must be seen in selections, because one could see it in its entirety only "by going to the moon and then somewhat obscurely; we see as much of it as we can get hold of." In 1903 travel to the moon was a matter of science fiction. The astronauts who first visited it in 1969 certainly echoed Chesterton in finding the sight of the earth in its entirety a most exciting view. But in 1903 even the boldest science fiction writers would not have suggested that it would ever be possible to do what Chesterton stated as impossible in the next breath: "The universe itself cannot show its unity; we have to judge it in selections."[44]

Fourteen years later, in 1917, Einstein showed in the last of his memoirs on general relativity that it was possible, if not to travel outside the universe and take a look at it, at least to account mentally and exactly for the totality of gravitational interactions. Gravitation, insofar as science deals with it quantitatively, is not matter or reality as such; it is only an aspect of material being. As it turned out, that aspect is really an aspect, that is, specificity. The particular space-time manifold, which is needed to deal with that quantitative aspect, is indeed very specific. And so is that number which gives the curvature of space-time produced by the total mass, another specific quantity, computed from the average density of mass, or what is the same, the number of galaxies within a sufficiently large volume of space.

This was only the beginning of the greatest story in the history of science, a story of which the latest chapters and pages are being written at an ever more feverish pace.[45] As the Einsteinian approach unfolded more and more specifics about the universe, those specifics revealed more and more their mutual coherence. First came the discovery, theoretical and empirical, of the expansion of the universe. There followed the connecting of the genesis of elements with the genesis of the universe. Scientific minds tired of, or simply not tuned to the stark specificity of things around them, are now confronted with a cosmic specificity which defies imagination, and are seized by a gripping sensitivity for metaphysics. The result was prophetically portrayed in Chesterton's final explanation of the optimism of Dickens's world view, tragic as the world could appear in his novels. From the remark that the ultimate basis of that optimism was Dickens's perception of the strangeness of every person and situation, Chesterton went on to fathoming the very depths of metaphysics: "But when all is said, as I have remarked before, the chief fountain in Dickens of what I have called cheerfulness, and some prefer to call optimism, is something deeper than a verbal philosophy. It is after all, an incomparable hunger and pleasure for the vitality and the variety, for the infinite eccentricity of existence. And this word 'eccentricity' brings us perhaps, nearer to the matter than any other. It is, perhaps, the strongest mark of the divinity of man that he talks of this world as 'a strange world,' though he has seen no other. We feel that all there is is eccentric, though we do not know what is the centre. This sentiment of the grotesqueness of the universe ran through Dickens' brain and body like the mad blood of the elves."[46]

Some very similar throbbing seems to have taken hold over the intellectual veins of scientific cosmologists of our times. It is indeed very difficult not to be struck by the thought that the evolution of the entire universe and of everything in it would have been very different if the primeval cosmic soup had not contained almost exactly 40 million photons for each electron, proton, and neutron. Such a strangely specific state of affairs, which clearly could have been otherwise, had, however, to be on hand, if the actual world corresponding to the Mendeleev table of elements was to emerge at the end of the so-called first three minutes.[47] In their exploration of the earliest phases of those first three minutes, modern scientific cosmologists noted several similarly baffling specificities. They all are so cunningly off-center, that is, so subtly deviating from the presumed perfect balance of homogeneity,[48] as to make their students dizzy with metaphysics to the point of appearing almost eccentric. All those "cosmic imbalances" attest that the universe is a most specific entity limited to a relatively few possibilities. Even more important, over such a universe there looms the enormously high probability of its not being at all in the very form as we know it.

The proportion of actually existing things, however numerous, to the true infinity of possible things should seem staggeringly small. The truth of this will stagger only if expressed in that staggering style of which Chesterton was a past master. He did not fail us even in this very connection. In fact he almost surpassed his very superlative expressiveness as he compared the things composing the actual universe to the list of the few items which Robinson Crusoe saved from the shipwreck: "It is a good exercise, in empty or ugly hours of the day, to look at anything, the coal-scuttle or the bookcase and think how happy one could be to have brought it out of the sinking ship to the solitary island. But it is a better exercise still to remember how all things have had this hair breadth escape: everything has been saved from a wreck. . . . It was common to say that many a man was a Great Might Have Been. To me it is a more solid and startling fact that any man in the street is a Great Might Not Have Been."[49]

On facing up to such a passage one cannot help feeling that Lawrence Clipper did not exaggerate in saying that "Chesterton's knack for finding the right example, the right analogy to make an abstract point is unique in the English letters."[50] Nor would one exaggerate in saying that the main message of modern scientific cosmology would never obtain an expression more comprehensible and more staggeringly to the point than that just quoted passage of Chesterton. It is the point of a radical, all pervading cosmic contingency. Only by keeping in view the handful of items, be they the list of Robinson Crusoe's scant possessions, or the entire inventory of the universe, shall one see the profundity of Chesterton's view that the "moment we have a view of the universe, we possess it."[51] This statement of his is perfectly applicable to what has come about through modern scientific cosmology, which, by giving us a scientifically ever more reliable view of the structure and shape of the universe, allows the universe to become man's finest possession.

Chesterton most likely referred to fairly well publicized debates raging around 1930 or so about the latest in scientific cosmology and to some skeptical views according to which, as he put it, "one can never find out the shape of the universe." Not that cosmology as such was his concern. He disputed moral relativism which refuses truth on the ground that opinions are varied and manifold. Such a position was no more reasonable in Chesterton's eyes than the claim that just because some people thought the earth to be flat, while others held it to be round, one was "free to say that it is triangular or hexagonal, or a rhomboid," or that "it has no shape at all or its shape can never be discovered," or that "anyhow, modern science must be wrong in saying it is an oblate spheroid." Whether Chesterton read about a spheroidal universe in Eddington's books or in the *Times* is irrelevant. While that idea was public property, Chesterton was alone with his emphatic assertion, one of the best contributions ever made to cosmology, both philosophical and scientific: "The world must be some shape, and it must be that shape and no other; and it is not self-evident that nobody can possibly hit on the right one."[52]

The story of scientific cosmology since 1932, when Chesterton made that brilliant statement, had indeed been a quest for an ever firmer grip on the true shape of the universe. One aspect of that story is particularly Chestertonian, inasmuch as it now seems that the universe was structured from the start in a way so that ultimately man might dwell in it. Scientific cosmologists have been calling this for over a decade now, the anthropic principle.[53] Chesterton in turn would say to them: "I kept telling you long ago and all the time that the universe was cozy."[54] Coziness could not have turned out to be a more cosmic quality.

It should remain the privilege of literary experts on Chesterton to speculate about the diminutives he would use today on hearing about the earliest phases of the universe. There the universe is rapidly shrinking as one traces time backward. From a universe shrunk to the size of a star the distance is only a few seconds to a universe shrunk to the size of planet, and from there the distance is measured in microseconds to a universe not larger than a pea or a pinhead. Chesterton would now readily qualify his erstwhile view that we cannot say whether the universe is young or old. He would not read scientific prophecy into his remark, made at least a decade before the big-bang theory had begun to be spoken of, that he had after all "often felt the physical universe as something like a firework display."[55] But on learning about the erstwhile pinhead-size of the universe—an infinitely far cry from an allegedly infinite homogeneous proto-universe—he could rightly claim the status of a prophet.

He would remind us with his jovially humble boastfulness that poets extracted themselves from *that infinity* long before scientists heard from Eddington that in science, and especially in cosmology, infinity was the real mischief maker.[56] This is not necessarily true of non-Euclidean infinities, but they are strange infinities indeed. One of their kind, the hyperbolic space-time-manifold, is no less strange than a saddle with very specifically bending slopes but with no edges. Such an infinity is a perfect example of that cosmos where "incompleteness implies completeness," to recall a priceless phrase from Chesterton's book on Robert Browning.[57] For Chesterton saw in this cosmic view the essence of Browning's philosophy. In praising it so highly he also declared it to be his own. And this is also the essence of modern scientific cosmology, whose best students found that in the actual physical world specificity (which is always incompleteness) is the hallmark not only of each and every part but also of their totality, the universe. Chesterton would also remind us that good philosophers had always known what scientists lately have discovered, namely, that "at the back of everything, existence begins with a harmony and not a chaos."[58] And on seeing the embarrassment of agnostic scientists, who cannot help seeing in the specific harmony of gluons, neutrons, photons, and other subparticles a pointer to a choice that only a Creator could make, he would recall a phrase from *Orthodoxy:* "Modern science moves toward the supernatural with the rapidity of a railway train."[59]

The comparison was timely. In 1908 giant locomotives had their golden age. Airplanes were yet a mere curiosity. Only some visionaries, like Tchiolkovsky and Oberth, made serious studies of space travel. A generation later, Chesterton himself became a space-age prophet by noting that if space travel were ever to become a reality "there will be a traffic problem about flying ships, exactly as there is now a traffic problem about taxicabs."[60] At any rate, Chesterton might refer today to the speed of light as if to indicate that science, insofar as all science is cosmology, has arrived at the point where it has to face up to the supernatural, which is the true Infinite, the Creator of all. Of course, from primitive archaic times on, the universe has always looked specific enough to invite the consideration about a cosmic choice determining the actual shape of things. Whatever true progress has been made in the history of science, it was always an advance from one stage of specificity to a stage where things appeared even more specific,[61] that is, ever more incomplete in their ever greater completeness. But, as I noted earlier, only since Einstein has science achieved a contradiction-free discourse about the totality of consistently interacting things, and in doing so it revealed a most specific universe. It is in that sense that science can be seen as carrying on with the speed of light to the supernatural and touching on it as does a champion on the finish line. The exact shape of that line will see many further refinements, but they all will bear further witness to a most specific cosmos, which is therefore radically contingent on a supracosmic choice for its existence.

To acknowledge the contingency of the universe is hardly a natural move. It has never been natural for fallen man to fall on his knees. Science, or rather the so-called scientific establishment and its pseudo-philosophical consensus, will keep itself light-years removed from the point where scientific cosmology readily becomes metaphysical cosmology and natural theology. Every trick of the trade—from multiworlds to cosmic quantum flips—is being tried out so that the metaphysical sting may disappear from modern scientific cosmology.[62] Most leading scientific cosmologists swear by the universe only to discredit that outlook on it which Chesterton celebrated under the caption: "The Flag of the World." Theirs is that old pagan view that makes God part of the universe and then turns Him into the universe itself. That today there are self-styled Christian theologians who do the same would not surprise Chesterton. Rather they, overawed as they are by an unjustified sense of originality, would be surprised on finding Chesterton decry a phenomenon very noticeable in the first decade of this century, the first heyday of modernism. In speaking of the Christian answer to the pessimism of pantheism, Chesterton defined it as the answer "which was like the slash of a sword; it sundered; it did not in any sense sentimentally unite. Briefly, it divided God from the cosmos." And he added: "That transcendence and distinctness of the deity which some Christians now want to remove from Christianity, was really the only reason why any one wanted to be a Christian. It was the whole point of the Christian answer to the unhappy pessimist and the still more unhappy optimist."[63]

If this was true, its contrary had to be no less valid, for, as Chesterton aptly put it, "religion means something that commits man to some doctrine about the universe."[64] Indeed, as history shows, all religions do. Chesterton was overjoyed on finding in Christopher Dawson's *Progress and Religion* a brief report of Duhem's monumental interpretation of the history of cosmology.[65] The gist of that interpretation consists in putting all religions in two categories. In one there is the Judeo-Christian religion with its belief in a linear cosmic story running from "In the beginning" to a "new heaven and earth." In the other are all the pagan religions, primitive and sophisticated, old and modern, which invariably posit the cyclic and eternal recurrence of all, or rather the confining of all into an eternal treadmill, the most effective generator of the feeling of unhappiness and haplessness. About that treadmill, usually spoken of as the doctrine of the Great Year, Chesterton's remark was: "I am exceedingly proud to observe that it was before the coming of Christianity that it flourished and after the neglect of Christianity that it returned."[66] Chesterton was fully aware of the inner logic of the wheel, the chief symbol of Buddhism, which he saw subtly return in Herbert Spencer's "cosmic conservation and recurrence."[67] Against it he found, as did St. Augustine against the doctrine of the Great Year,[68] in the unrepeatable resurrection of Christ the only effective antidote: "For it is the key to nearly everything in the development of two thousand years; and in nothing is it so much the key to Christendom as in the recurrent reversions to Paganism."[69]

Whatever the lacunae, from a scientific viewpoint, in Chesterton's dicta on the cosmos, they remain very prophetic and profound by any comparison. Such had to be the case, since he grasped the basis of the only good philosophy which is a commitment to things in all their actual variety and specificity. Therefore his philosophy had to become a cosmic philosophy, a fact of which he was utterly conscious. He stated that unless philosophy was cosmic and eternal it was not philosophy.[70] No less important, he saw that just as the student of origins can make only one mistake,[71] the cosmic philosopher is no better off. His possibly most fatal mistake is to choose as his starting point not the things but the self or even the phenomena, insofar as they are his mere sensations: "A cosmic philosophy is not constructed to fit man; a cosmic philosophy is constructed to fit a cosmos. A man can no more possess a private religion than he can possess a private sun or moon."[72] No wonder that such an insightful and consistent thinker spoke devastating words of that philosopher, Kant, who more than any other succeeded in leading mankind into the belief that the universe was the bastard product of the metaphysical cravings of the intellect. Long essays on Kant and the German idealists contain far less than these few words of Chesterton: "The note of our age is a note of interrogation. And the final point is so plain; no sceptical philosopher can ask any questions that may not equally be asked by a tired child on a hot afternoon. 'Am I a boy?—Why am I a boy?—Why aren't I a chair?— What is a chair?' A child will sometimes ask questions of this sort for two hours. And the philosophers of Protestant Europe have asked them for two hundred years."[73] By asking to no end such questions, and even more stupefying ones, Kant and his camp wanted to secure the independence of the self from reality and ultimately from that reality which is God.[74] Nothing can indeed make belief in God more illusory than a studied disbelief in any and all reality which make up the universe. Chesterton saw fully through the self-defeating nature of that strategy as he insisted that if there was no universe, there was nothing to think of, not even a philosopher, let alone God.[75]

To those among modern scientific cosmologists who today boast of their solipsism Chesterton would say: "The colour-blind man may rejoice in the fairy-trick which enables him to live under a green sun and blue moon. But if once it be held that there is nothing but a silver blur in one man's eye or a bright circle (like a monocle) in the other man's, then neither is free, for each is shut up in the cell of a separate universe."[76] As to those scientific entertainers who on the TV screen clothe the universe in flat witticisms so that neither man nor God may transpire, he would tell with contempt what he said about H. G. Wells: "He began by making a new heaven and a new earth, with the same irresponsible instinct by which men buy a new necktie or button-hole. He began by trifling with the stars and systems in order to make ephemeral anecdotes; he killed the universe for a joke."[77] In view of the vast sums of money earned by such jokes, Chesterton might speak of killing the universe for sumptuous royalties.

Such phrases of Chesterton are sparkling gems. But if there is to be a revival of interest in Chesterton, such gems will not have the sufficient attractiveness to bring it about. Attractiveness is a correlation between object and subject. Unless the subject has a hunger for things, it will look for a sparkle not in them but in the self, that possibly dullest thing. Chesterton did speak very attractively. Had he done nothing else he would have become one of those transitory greats of style of whom the history of letters recorded many but finds perennially fresh hardly one. Chesterton not only spoke attractively, but he did so of a vast variety things. Prominently among those things was science, the most stupendous thing of modern life and the most inexhaustible source of ever fresh things. There are several excellent ways of preventing that stupendous thing from turning into a stupefying monster. One of them is to discover in Chesterton's words on science the words of a seer.

List of Chesterton's Books Quoted

All I Survey (London: Methuen, 1933)

All Is Grist (New York: Dodd, Mead, 1932)

All Things Considered (New York: John Lane, 1909)

As I Was Saying (New York: Dodd, Mead, 1936)

Autobiography (London: Hutchinson, 1936)

Avowals and Denials (New York: Dodd, Mead, 1935)

The Ball and the Cross (New York: John Lane, 1909)

The Catholic Church and Conversion (New York: Macmillan, 1926)

Charles Dickens: A Critical Study (New York: Dodd, Mead, 1906)

The Club of Queer Trades (London: Harper & Brothers, 1905)

Come to Think of It (New York: Dodd, Mead, 1931)

The Common Man (New York: Sheed and Ward, 1950)

The Defendant (New York: Dodd, Mead, 1904)

Eugenics and Other Evils (London: Cassell, 1922)

The Everlasting Man (New York: Dodd, Mead, 1925)

Fancies versus Fads (New York: Dodd, Mead, 1923)

G. F. Watts (London: Duckworth, 1904)

G. K. C. as M. C. (London: Methuen, 1929)

Generally Speaking (New York: Dodd, Mead, 1929)

George Bernard Shaw (New York: John Lane, 1910)

The Glass Walking Stick and Other Essays (London: Methuen, 1955)

Heretics (New York: John Lane, 1905)

Lunacy and Letters (New York: Sheed and Ward, 1958)

The Man Who Was Thursday (1908; New York: Penguin Books, 1976)

Manalive (1912; Philadelphia: Dufour, 1962)

A Miscellany of Men (London: Methuen, 1912)

The New Jerusalem (London: Hodder and Stoughton, 1921)

Orthodoxy (London: John Lane, 1909)

The Outline of Sanity (New York: Dodd, Mead, 1927)

The Poet and the Lunatics (New York: Dodd, Mead, 1929)

The Resurrection of Rome (New York: Dodd, Mead, 1930)

Robert Browning (1906; New York: Macmillan, 1916)

St. Thomas Aquinas (New York: Sheed and Ward, 1933)

The Spice of Life and Other Essays (Beaconsfield: Darwen Finlayson, 1964)

The Thing (London: Sheed and Ward, 1929)

Tremendous Trifles (1909; New York: Sheed and Ward, 1955)

The Uses of Diversity (New York: Dodd, Mead, 1921)

The Well and the Shallows (New York: Sheed and Ward, 1935)

What I Saw in America (London: Hodder and Stoughton, 1922)

What's Wrong with the World (New York: Dodd, Mead, 1910)

Notes

CHAPTER ONE: INTERPRETER OF SCIENCE

1. *What's Wrong with the World*, p. 320. (A list of books by Chesterton quoted in this volume is given in the preceding two pages; publishers and places and dates of publication are included there. Citations to Chesterton's books throughout the notes will be by title and page only.)

2. [Frances B. Chesterton,] *Wit and Wisdom of G. K. Chesterton* (New York: Dodd, Mead, 1911). In the five hundred or so passages contained in that book only half a dozen are directly related to science, which is mentioned incidentally in another half a dozen.

3. [Cecil Chesterton,] *Gilbert K. Chesterton: A Criticism* (New York: John Lane, 1909), p. 125.

4. "Walking Tours," *Daily News* (London), 23 Sept. 1901; see *G.K. Chesterton: The Apostle and the Wild Ducks and Other Essays*, ed. Dorothy E. Collins (London: Paul Elek, 1975), p. 57.

5. *Heretics*, p. 171.

6. *G.F. Watts*, p. 120.

7. Such as Shaw, Kipling, Moore, Dickinson, and Whistler.

8. [Cecil Chesterton,] *Gilbert K. Chesterton*, pp. 124–25.

9. Maisie Ward, *Gilbert Keith Chesterton* (London: Sheed and Ward, 1944), p. 33.

10. *Autobiography*, p. 52.

11. Ibid., pp. 105-6.

12. *Fancies versus Fads*, p. 217.

13. Maisie Ward, *Return to Chesterton* (London: Sheed and Ward, 1952), p. 9.

14. *Culture and the Coming Peril: Being the Seventh of a Series of Centenary Addresses* (London: University of London Press, 1927), p. 5.

15. *Autobiography*, p. 105.

16. *The Catholic Church and Conversion*, p. 77.

17. *Autobiography*, p. 106.

18. For a criticism of this and other similar contentions made by Snow in his *Two Cultures,* see my *Culture and Science: Two Lectures Delivered at Assumption University, Windsor, Canada, on February 26 and 28, 1975* (University of Windsor Press, 1975), pp. 14-20; reprinted from the *University of Windsor Review* (Fall 1975), pp. 55-104.

19. Julius West, *G.K. Chesterton: A Critical Study* (London: Martin Secker, 1915), pp. 42, 44, 56.

20. Alan Handsacre [A. C. White], *Authordoxy: Being a Discursive Examination of Mr. G. K. Chesterton's "Orthodoxy"* (London: John Lane, 1921).

21. In his *The Innocence of G.K. Chesterton* (London: Cecil Palmer, 1923) Gerald William Bullett, an avowed agnostic, approvingly states about Chesterton that "he derides, as who does not, the pretensions of scientists to answer the unanswerable enigma." Yet, Bullett held that there would eventually be a natural (scientific) explanation for even those most extraordinary phenomena which believers take for miracles (pp. 43, 91). Bullett wholly missed the bearing for the understanding of science of *The Ball and the Cross* and of "The Ethics of Elfland."

22. Joseph de Tonquédec, *G. K. Chesterton, ses idées et son caractère* (Paris: Nouvelle Librairie Nationale, 1920).

23. W. F. R. Hardie, "The Philosophy of G. K. Chesterton," *Hibbert Journal* 29 (1931): 449–64; see especially pp. 452–53.

24. Patrick Braybrooke, *Gilbert Keith Chesterton* (Philadelphia: J.B. Lippincott, 1922), *A Chesterton Catholic Anthology* (London: Burns, Oates & Washbourne, 1928), and *The Wisdom of G.K. Chesterton* (London: Cecil Palmer, 1929). The little which Braybrooke says in the latter on science in ch. 3, "The Essayist," relates entirely to H. G. Wells (pp. 105–6). There are no references whatsoever to science in Braybrooke's discussion of *Orthodoxy* where he passes over in silence "The Ethics of Elfland" (see ch. 7, "The Christian.") Science is no less ignored in George Schuster's "The Adventures of a Journalist: G. K. Chesterton," ch. 13 in *The Catholic Spirit in Modern English Literature* (New York: Macmillan, 1922).

25. Chesterton gave that estimate in a letter of July 3, 1909, to Father O'Connor himself. See John O'Connor, *Father Brown on Chesterton* (London: Frederick Muller, 1938, p. 123). The critique of *Orthodoxy* quoted there (pp. 164–66) appeared in the October 1, 1908, issue of the *Times Literary Supplement*.

26. Maurice Evans, *G. K. Chesterton: The Le Bas Prize Essay 1938* (Cambridge: University Press, 1938).

27. Frank Alfred Lea, *The Wild Knight of Battersea: G. K. Chesterton* (London: James Clarke: 1945).

28. Lawrence J. Clipper, *G. K. Chesterton* (New York: Twayne, 1974), p. 184.

29. Lea, *The Wild Knight of Battersea*, p. 49.

30. *Orthodoxy*, p. 154.

31. Charles Bradlaugh, *A Plea for Atheism* (London: Freethought Publishing Co., 1877), p. 10.

32. Hugh Kenner, *Paradox in Chesterton* (New York: Sheed & Ward, 1947).

33. Christopher Hollis, *The Mind of Chesterton* (London:

Hollis and Carter, 1970); see especially pp. 272–82. In discussing *Orthodoxy* (pp. 71–75) Hollis made no reference to "The Ethics of Elfland," nor did he speak of Chesterton's notion of the universe as he discussed *Heretics*.

34. Sister M. Carol, *G. K. Chesterton: The Dynamic Classicist* (Delhi: Motilal Banarsiddas, 1971).

35. Gary Wills, *Chesterton: Man and Mask* (New York: Sheed & Ward, 1961). The point is clearly conveyed by a glance at the combined name and subject index which does not contain such entries as science or evolution. Science is briefly mentioned by Wills as he presents the argument of *Heretics:* a defense of the particular against the general to which, according to Wills, science reduces its subject matter (see p. 88).

36. That book, A. D. Nuttal's *A Common Sky: Philosophy and the Literary Imagination* (Berkeley: University of California Press, 1974), was dedicated by its author to Stephen Medcalf, whose essay, "The Achievement of G. K. Chesterton," appeared in John Sullivan, ed., *G. K. Chesterton: A Centenary Appraisal* (New York: Harper & Row, 1974), pp. 81–121; for quotation see p. 116. The remaining twelve essays in Sullivan's book are wholly irrelevant to our topic. Ten or so years before Medcalf, Alan L. Maycock too left unexploited the subject of creation, although he reprinted in part in his *The Man Who Was Orthodox: A Selection from the Uncollected Writings of G. K. Chesterton* (London: Dennis Dobson, 1963) a powerful early essay of Chesterton dealing with pantheism versus creation, which I discuss in chapter four.

37. Clipper, *G. K. Chesterton*, pp. 117, 133. Science was nowhere in sight as Clipper discussed "The Ethics of Elfland."

38. Such a portrait is *G. K. Chesterton: Radical Populist* by Margaret Canovan (New York: Harcourt, Brace, Jovanovich, 1977).

39. W. H. Auden, ed., *G. K. Chesterton: A Selection from His Non-fictional Prose*, (London: Faber & Faber, 1970), p. 17.

40. Ibid., p. 18. In addition to "The Ethics of Elfland" three other selections, out of a total of thirty-six, related to science. Two of them, a single page each, were from *Orthodoxy* and related respectively to Darwinism and morality, and to nature and logic. Auden reproduced in full ch. 11, "Science and the Savages," from *Heretics*. All this was hardly a great improvement on *G. K. Chesterton: An Anthology*, selected with an introduction by D. B. Wyndham Lewis (London: Oxford University Press, 1957), in which even Chesterton the philosopher was largely missing.

41. Lynette Hunter, *G. K. Chesterton: Exploration in Allegory* (New York: St. Martin's Press, 1979).

42. Ian Boyd, *The Novels of G. K. Chesterton: A Study in Art and Propaganda* (New York: Harper & Row, 1975); see especially pp. 20-32, where even the word *scientist* appears only fleetingly.

43. Alzina S. Dale, *The Outline of Sanity: A Life of G. K. Chesterton* (Grand Rapids, Mich.: Eerdmans, 1982), p. 291. Dale, in claiming, as she discussed Chesterton's apologetics, that Chesterton was "torn between feeling and reason, because reason, too, is fallacious and must be completed by faith," seems to have missed the powerfully realist philosopher in Chesterton, who only on the rarest occasions spoke not as a Thomist but as an apparent fideist for whom the grasp of reality is not a question of knowledge but of faith. Even then, as a comparison of pp. 58 and 61 in *Orthodoxy* shows, Chesterton quickly reverts to that realism which is a thorough trust in the mind's ability to reach objectively existing reality.

44. The essays in question are J. Gordon Parr, "Chesterton and Technology," and J. G. Keogh, "Chesterton: Technology,

Culture and Anarchy," both in *Chesterton Review* 3 (1976), 91–98 and 287–92.

45. Cyril Clemens, *Chesterton as Seen by His Contemporaries*, with an introduction by Edmund C. Bentley (Webster Groves, Mo.: International Mark Twain Society, 1939).

46. Joseph W. Sprug, ed., *An Index to G. K. Chesterton*, (Washington D.C.: Catholic University of America Press, 1966).

47. John Sullivan, *G. K. Chesterton: A Bibliography, with an Essay on Books by G. K. Chesterton and an Epitaph by Walter De La Mare* (London: University of London Press, 1958) and *Chesterton Continued: A Bibliographical Supplement* (London: University of London Press, 1968).

48. Martin Gardner, ed. *Great Essays in Science* (New York: Pocket Books Inc., 1957), pp. 78–83. Gardner kept the same selection with the same introduction to it in the revised edition published as *The Sacred Beetle and Other Great Essays in Science* (Buffalo: Prometheus Books, 1984), pp. 95–101.

49. *Great Essays in Science*, p. 77.

50. The selection runs in *Orthodoxy* from the middle of p. 87 to the middle of p. 97; unannotated quotations in the following pages are from that section. Two years earlier, in 1906, Chesterton had already celebrated Elfland in his *Charles Dickens:* "Every train of thought may end in an ecstasy and all roads lead to Elfland" (p. 20).

51. The brilliance of *Orthodoxy* is in part responsible for the neglect of much earlier though no less impressive statements of Chesterton on major philosophical points stressed in *Orthodoxy* and in particular in "The Ethics of Elfland." Thus in an essay "The Revival of Philosophy—Why?" the same argument is made in a no less memorable phrasing: "If a man sees a river run downhill day after day and year after year, he is justified in reckoning, we might say in betting, that it will do so till he dies.

But he is not justified in saying that it cannot run uphill, until he really knows why it runs downhill. To say it does so by gravitation answers the physical but not the philosophical question. It only repeats that there is a repetition; it does not touch the deeper question of whether that repetition could be altered by anything outside it. And that depends on whether there is anything outside it. Mere repetition does not prove reality or inevitability" (The Common Man, pp. 178–79). Neither this essay nor the other essays that form The Common Man are identified by original publication dates. Regardless of whether I am right in assigning that essay to around 1901 or perhaps earlier, Chesterton achieved very early in his career that momentous insight into repetitions in nature. I came across a proof of this several months after the delivery of these lectures, when I was invited to present a paper at the Chesterton Interdisciplinary Symposium at Catholic University, Washington, on March 10, 1984. In choosing for my topic the hitherto unexplored Chesterton-Blatchford controversy of 1903, described by Chesterton in his Autobiography as "The Landmark Year" in his life, I was led to Chesterton's essays buried in The Clarion (Blatchford's weekly). In order to disprove Blatchford's scientism, a chief aim of which was the discrediting of miracles, Chesterton focused on the repetitions assumed by those formulating scientific "laws": "The question of miracles is merely this: Do you know why a pumpkin goes on being a pumpkin? If you do not, you cannot possibly tell whether a pumpkin could turn into a coach or couldn't. That is all. All the other scientific expressions you are in the habit of using at breakfast are words and winds." About the same time, and again in The Clarion, Chesterton made an unconditional declaration of his belief in the divinity of Christ, as a belief which alone can safeguard human sanity. For further details, see my article, "Chesterton's Landmark Year, or the

Blatchford-Chesterton Debate of 1903–04," published in *Chesterton Review* 10 (1984): 409–23.

52. As the author of a book on David Hume, Huxley should have been more cautious, but his was the firm belief that in principle the scientific method enables the scientist to predict the exact trajectory of any and all vapor molecule forming a mist. (Huxley made this particular point in his reminiscences on the reception of Darwin's *Origin;* see *The Life and Letters of Charles Darwin,* ed. Francis Darwin (London: John Murray, 1887, vol. 2, p. 200; rpt. New York: Basic Books, 1959). Huxley's simile was a picturesque elaboration on the manner in which Laplace had spoken two generations earlier of the predictive powers of a superior spirit to whom all actual parameters of all material bodies are fully known.

53. Bertrand Russell, *Philosophy* (New York: W. W. Norton, 1927), p. 14.

54. The point at issue was the "selective indignation" inherent in most pacifist philosophies: "Mr. H. G. Wells has written a thousand pages in favour of Peace, but not one page in favour of Poland. Lord Russell has said much, from his point of view, to deter men from fighting, but nothing that would deter Mussolini from fighting; and nothing certainly that could deter any Communist from fighting Mussolini. To examine, prove, disprove, or reconcile the philosophies of Europe—that would be a task for a philosopher, but not for a philosopher like Bertrand Russell. That is the only way to Peace; and few be they that find it." *Avowals and Denials,* pp. 229–30.

55. *Orthodoxy,* p. 167.

56. *St. Thomas Aquinas,* p. 176.

57. In the same statement, which became part of the Chesterton literature through Clemens's *Chesterton as Seen by His Contemporaries* (pp. 150–51), Gilson also stated: "Nothing short of

genius can account for such an achievement. . . . He has guessed all that which we had tried to demonstrate, and he has said all that which they were more or less clumsily attempting to express in academic formulas. Chesterton was one of the deepest thinkers who ever existed; he was deep because he was right; and he could not help being right; but he could not either help being modest and charitable, so he left it to those who could understand him to know that he was right, and deep; to others, he apologized for being right, and he made up for being deep by being witty. That is all they can see of him."

58. Unfortunately, this letter of Gilson, of which a typewritten transcript was kindly put at my perusal by the Rev. Ian Boyd, is still to appear in full in print. The letter was written on January 7, 1966, to the Rev. Kevin Scannell. Its content and tone clearly refutes those who, like Wills (*Chesterton*, pp. 4, 179) and Clipper (*G. K. Chesterton*, p. 111), felt that Gilson's words did not represent his most considered judgement on Chesterton's stature as a philosopher. In speaking of Chesterton the philosopher it may also be noted that contrary to Wills (p. 4) Kenner nowhere stated in his *Paradox in Chesterton* that Chesterton was only a philosopher and not also a journalist and a debater. Kenner merely stated (pp. 6-7, 25-26) that not only was philosophy always at the basis of any discourse, be it that of a journalist, but that in Chesterton's case the only sound philosophy was at the basis and in a superbly sparkling form at that. Particularly instructive should seem in this respect Chesterton's essay, "The Revival of Philosophy—Why?," quoted in note 51 above.

59. *Orthodoxy*, p. 97. Subsequent unannotated quotations are from the remainder of "The Ethics of Elfland" (pp. 97-116).

60. The italics merely serve the purpose of calling attention to the philosophical significance that Chesterton clearly seems to attach to the word *happen*. It is in fact the shortest though

rather worn-off expression of contingency. What is really and absolutely necessary never happens but IS.

61. An accusation made by Clipper, *G. K. Chesterton*, p. 154.

62. Pearson's exclusive attention to facts, an attention all too characteristic of the *Grammar of Science,* is the chief target of Chesterton's criticism of Pearson's advocacy of eugenics: "There is the other kind of man, like Dr. Karl Pearson, who undoubtedly knows a vast amount about his subject; who undoubtedly lives in great forests of facts concerning kinship and inheritance. But it is not, by any means, the same thing to have searched the forests and to have recognized the frontiers. Indeed, the two things generally belong to two very different types of mind" *(Eugenics and Other Evils,* p. 65).

63. This was in fact the theme on which Poincaré's *The Value of Science* came to a close: "All that is not thought is pure nothingness; since we can think only of thought and all the words we use to speak of things can express only thoughts, to say there is something other than thought, is therefore an affirmation which can have no meaning."

64. See the *Autobiography,* pp. 92–94. An illuminating contrast to Chesterton's reconquest of reality is the obsessive fear of anything objective we find in Sartre's *Nausea* or Rousseau's morbid celebration of non-being as the only "thing" which is "beautiful."

65. Hilaire Belloc, "Gilbert Keith Chesterton," *Saturday Review of Literature* 14 (July 4, 1936), 4

66. *The Poet and the Lunatics,* p. 124. To grasp the philosophical depth of that passage one is well advised to recall Gilson's pointed warning in his *The Unity of Philosophical Experience* that unlike the various branches of rationalism, idealism, sensationism, and positivism, Marxism is a serious philosophy on account of the realism inherent in its materialism.

67. Of the many passages in which Chesterton celebrates common sense, the one in which he describes its weakening through exaggerated attention to the scientific method is especially appropriate: "The unmistakable mood or note that I hear from Hanwell, I hear also from half the chairs of science and seats of learning today; and most of the mad doctors are mad doctors in more sense than one. They all have exactly that combination we have noted: the combination of an expansive and exhaustive reason with a contracted common sense. They are universal only in the sense that they take one thin explanation and carry it very far" (*Orthodoxy,* p. 36).

68. For details, see my *Uneasy Genius: The Life and Work of Pierre Duhem,* (Dordrecht: Martinus Nijhoff, 1984), pp. 368–71.

69. The argument of Thomas S. Kuhn's book, *The Structure of Scientific Revolutions* (Chicago: University of Chicago Press, 1962) and the debates it stirred are discussed and documented in ch. 15, "Paradigms or Paradigm," of my Gifford Lectures, *The Road of Science and the Ways to God* (Chicago: University of Chicago Press, 1978).

70. *Orthodoxy,* pp. 74, 71.

71. This is, of course, true, as Aristotle himself points it out, only as long as the dictator (tyrant) lives up to the tactic which secures his position. See *Politics,* Bk. 5, chs. 10 and 12.

72. Gilson reported that when pressed for explicit statements on that point, Meyerson usually replied with an enigmatic smile. E. Gilson, T. Langan and A. A. Maurer, *Recent Philosophy: Hegel to the Present* (New York: Random House, 1966), p. 289.

73. An approach initiated by Bachelard and fully articulated by Koyré and his disciples. For details see my *The Road of Science and the Ways to God,* ch. 15.

74. Albert Einstein, letter of December 22, 1950, to E. Schrö-

dinger; see *Letters on Wave Mechanics: Schrödinger, Planck, Einstein, Lorentz*, ed. Karl Przibram, translated with an introduction by M. J. Klein (New York: Philosophical Library, 1967), p. 36.

75. For documentation and discussion, see my "Chance or Reality: Interaction in Nature versus Measurement in Physics," *Philosophia* (Athens) 10-11 (1980-81): 85-105.

76. *Orthodoxy*, p. 34.

CHAPTER TWO: ANTAGONIST OF SCIENTISM

1. *Orthodoxy*, p. 110.

2. The standard account of that hold is *Social Darwinism in American Thought* (1944; rev. ed. Boston: Beacon Press, 1955) by R. Hofstadter.

3. Spencer's encomium of science is all the more noteworthy as it comes to a close not merely with the declaration that "necessary and eternal as are its truths, all Science concerns all mankind for all time," but with the additional prophecy (logically implied in scientism) that other branches of learning will, like the "haughty sisters of an Eastern fable sink into merited neglect," whereas "Science, proclaimed as highest alike in worth and beauty, will reign supreme." Herbert Spencer, *Education: Intellectual, Moral and Physical* (1860; New York: D. Appleton, 1889), pp. 94, 96.

4. *Orthodoxy*, p. 109.

5. *The Well and the Shallows*, p. 57.

6. As stated by Darwin himself at the outset of his *Descent of Man* (new ed.; London: John Murray, 1901), p. 4.

7. *The Well and the Shallows*, p. 56.

8. *All Is Grist*, p. 60.

9. See Richard H. Costa, *H. G. Wells* (New York: Twayne, 1967), p.31.

10. In his famed Romanes Lecture, "Evolution and Ethics" (1893), Huxley also deplored the "moral flavour" attached by many evolutionists to the "survival of the fittest" and declared: "Let us understand, once and for all, that the ethical progress of society depends, not on imitating the cosmic process, still less in running away from it, but in combating it." See *Evolution Ethics and Other Essays* (New York: D. Appleton, 1914), p. 83. Admiration for Huxley's unflinching logic should not, however, blind one to the failure of Huxley, or of any evolutionist (or rather Darwinist), to explain how a most unethical and remorseless struggle for life could give rise to a species (Homo sapiens) which must submit to ethics in order to survive.

11. Quoted in Costa, *H. G. Wells*, p. 39.

12. H. G. Wells, *Seven Science Fiction Novels* (New York: Dover, 1934), p. ix. There Wells also admits that "I did my best to express my vision of the aimless torture in creation."

13. *Orthodoxy*, p. 109.

14. That it was hopeless "to trace a [cosmic] pattern of any sort" and that it was impossible to resist the conclusion that "homo sapiens . . . is in his present form played out" were the final messages of science as perceived by Wells in *Mind at the End of Its Tether* (New York: Didier, 1946), pp. 17, 18, 34. The only ray of comfort in that disconsolate picture was for Wells the assurance that at least a very small minority of mankind will succeed with Stoic resignation "in seeing life out to its inevitable end." One wonders whether Wells in writing his intellectual epitaph remembered what Chesterton had written only ten years earlier, and very prophetically, of Wells's scientific Utopianism resting on the claim that "intelligence cannot ulti-

131

mately be defeated." Noted Chesterton, "I might say, that I see no purely rationalist proof that intelligence cannot be defeated; ... men like Mr. Wells did talk as if Progress would be so intelligent as to relieve us of one problem after another; and did not allow enough for the fact that Progress itself might add yet another problem." *As I Was Saying,* p. 28.

15. *All Is Grist,* pp. 130-31.

16. *What's Wrong with the World,* p. 24.

17. *The Uses of Diversity,* pp. 137-38.

18. Forel's 62-page booklet (London: The New Age Press, 1908), translated from the German from a much publicized lecture he had given in 1906, was a summary of an almost ten times longer work by Forel, *The Sexual Question: A Scientific, Psychological, Hygienic and Sociological Study for the Cultured Classes* (English adaptation by C. F. Marshall), which saw a dozen re-editions between 1907 and 1941, to say nothing of the many printings of the German original and other translations. Beneath the highly elevated tone (absent only when it came to Catholic teaching on sexual ethics) there lay the kind of purely utilitarian thinking according to which "health" was the supreme norm, for the sake of which strict moral laws were to be disregarded to at least a "moderate" degree. Chesterton's ire was aroused precisely by that flouting of logic.

19. The decrease in number of families is a main concern for Amitai Etzioni in *An Immodest Agenda: Rebuilding America before the 21st Century* (New York: McGraw Hill, 1982). According to statistics presented by Etzioni, there would be no married Americans by 2008 if the over 1% annual rate of decrease of their numbers for the past 15 years were to remain constant. In his plea for stable families, which he calls "nuclear families" in order not to appear endorsing genuinely "conservative" views,

Etzioni displays the kind of evasiveness that would be a most welcome target for Chesterton's pen.

20. The rallying of *Time* (Jan. 17, 1983, p. 47) on behalf of children, so many defenseless victims of pedophiles, to me strangely contrasts with its chronic insensitivity for the far more defenseless human embryos, slated by the millions to be aborted. *Time* (Dec. 13, 1982, p. 74) voiced only doubts about the advisability of the legalization of homosexual unions.

21. See *New York Times*, Jan. 15, 1983, p. 12. SETI's chief advocate, Carl Sagan, would do well to ponder a remark, now almost a quarter of a century old, of William W. Howells, president of the American Anthropological Association, who in his book, *Mankind in the Making* (Garden City, N.Y.: Doubleday, 1959, p. 345), wrote: "I will lay a small bet that the first men from Outer Space will be neither bipeds nor quadrupeds but bimanous quadrupedal hexapods. (I have just invented the last word, in the hope that it means six limbs.)" There is more respect for the inner logic of Darwinian evolutionism in this brief spoof of extraterrestial intelligence than in the prolific publications and splashy videotapes of its ardent defenders.

22. *Generally Speaking*, p. 14.

23. *The Thing*, p. 87.

24. *The Uses of Diversity*, p. 244.

25. *Tremendous Trifles*, p. 152.

26. *Orthodoxy*, p. 228.

27. *The Well and the Shallows*, p. 74.

28. *Come to Think of It*, p. 161.

29. *All I Survey*, pp. 81, 82.

30. *All Is Grist*, pp. 29-30.

31. *St. Thomas Aquinas*, p. 9.

32. At the meeting of April 6, 1922, of the *Société française*

de philosophie; for the text of statements made there, see the Société's *Bulletin* 17 (1922): 91–113. To the point made by Bergson that a careful study of relativity theory would vindicate the primacy of common sense even with respect to the validity of its witness to the simultaneity of events, Einstein replied with the words, "there is no such thing as a philosopher's time; there is only a psychological time different from the time of physicists" (p. 107), which shows that Einstein was at that time far from perceiving the full extent to which his scientific achievement implied philosophical realism.

33. *The New Jerusalem*, p. 179.

34. *The Resurrection of Rome*, p. 98. Chesterton could not learn from even the better popularizations of relativity theory, let alone from its journalistic accounts, that in relativity theory "space" could only mean the network of permissible paths of motion and that its continued presence in Einstein's and other scientists' vocabulary was highly misleading. Chesterton would have had no objection to a statement such as that a material particle's path of motion was necessarily curved, however slightly, in the presence of another material particle.

35. *Come to Think of It*, p. 20.

36. *The Well and the Shallows*, pp. 56, 55.

37. *All Is Grist*, pp. 59–60.

38. *The Well and the Shallows*, pp. 57–58.

39. *All Is Grist*, p. 48.

40. *The Well and the Shallows*, pp. 56–57.

41. The age-old lure of apriorism is very much alive among leading physicists even today. It remains to be seen whether a better familiarity on their part with the implications of Gödel's incompleteness theorems would make them realize the utter futility of work on an all-encompassing physical theory which not only would fit the universe in all details but would do so with an a priori necessity. For further details, see my *Cosmos*

and Creator (Edinburgh: Scottish Academic Press; Chicago: Regnery-Gateway, 1980), pp. 49–54.

42. *All I Survey*, p. 102.

43. *All Things Considered*, pp. 226–27.

44. *Orthodoxy*, p. 235.

45. *All Things Considered*, p. 227.

46. Professor Heitler's statement, "In the decline of ethical standards, which the history of the past fifteen years exhibited, it is not difficult to trace the influence of mechanistic and deterministic concepts which have unconsciously, but deeply, crept into human minds," is noteworthy for two additional reasons: First, it refers to the period 1933–48; second, it is part of his contribution, "The Departure from Classical Thought in Modern Physics," to the famed volume, *Albert Einstein: Philosopher Scientist*, ed. P. A. Schilpp (Evanston, Ill.; Library of Living Philosophers, 1949), p. 196.

47. *All Things Considered*, p. 187. The same is to be said of Chesterton's parody of the opposite extreme of the scientific ideal, namely, sheer empiricism: "Science itself is only the exaggeration and specialization of this thirst for useless fact, which is the mark of the youth of man." *The Defendant*, p. 74.

48. *What's Wrong with the World*, p. 170.

49. *Fancies versus Fads*, p. 33.

50. *The Defendant*, p. 75.

51. A detailed discussion of the Arnold-Huxley belles-lettres versus science debate is given in my *Culture and Science: Two Lectures Delivered at Assumption University, Windsor, Canada, on February 26 and 28, 1975* (University of Windsor Press, 1975) pp. 3–12.

52. *The Everlasting Man*, p. 47.

53. "The Temple of Everything," *Daily News*, March 24, 1903, p. 8. In 1921 Chesterton spoke (*The New Jerusalem*, p. 174) of Huxley as "not only a man of genius in logic and

rhetoric" but also "a man of very manly and generous morality" who "deserves much more sympathy than many of the mystics who have supplanted him."

54. *The Club of Queer Trades*, p. 236. In the novel Grant is clearly Chesterton's mouthpiece.

55. *Orthodoxy*, p. 150.

56. For details, see my *Keys of the Kingdom: A Tool's Witness to Truth* (Chicago: Franciscan Herald Press, 1986).

57. "Since a babe was born in a manger, it may be doubted whether so great a thing has passed with so little stir." Alfred North Whitehead, *Science and the Modern World* (1925; New York: The New American Library of World Literature, 1948), p. 10.

58. *The Resurrection of Rome*, pp. 126-27.

59. "In these days we are accused of attacking science because we want it to be scientific," he wrote in *All Things Considered*, p. 187.

60. *Come to Think of It*, p. 163.

61. *Heretics*, p. 72.

62. Ibid., pp. 228, 229.

63. Jacob Bronowski, *Science and Human Values* (New York: Harper & Row, 1959), p. 90. Not surprisingly, such claim closely follows Bronowski's declaration: "This is the scientist's moral: that there is no distinction between ends and means" (ibid., p. 84).

64. *Come to Think of It*, p. 158.

65. *Eugenics and Other Evils*, "To the Reader." Unfortunately, many (better than tenth-rate) scientists are involved in that bureaucracy which calls for nothing less than Chesterton's stricture: "The thing that really is trying to tyrannise through government is Science. The thing that really does use the secular arm is Science." And since tyranny is always based on willfulness, Chesterton could rightly add: "The Inquisitor violently enforced his

creed, because it was unchangeable. The savant enforces it violently because he may change it the next day" (ibid. pp. 76, 78).

66. *What's Wrong with the World*, pp. 296-98.

67. *The Ball and the Cross*, pp. 280-82. It was these pages which prompted Julius West, insensitive to the enormous difference between science and scientism, to remark: "There follows a characteristic piece of that abuse [of science] for which Chesterton has never attempted to suggest a substitute." *G. K. Chesterton: A Critical Study* (London: Martin Secker, 1915), p. 42. The true merits of such judgement need not be discussed if a brief recall is made of Chesterton's encomiums of science quoted in the foregoing pages.

68. *In the Matter of J. Robert Oppenheimer*, (Washington D.C.: U.S. Government Printing Office, 1954), p. 81.

69. Herman Melville, *Moby Dick, or the Whale* (New York: Hendricks House, 1952), p. 185.

70. *Heretics*, p. 30.

71. For the first two of these statements of Einstein and for their contexts, see my *The Relevance of Physics*, (Chicago: University of Chicago Press, 1966), pp. 384, 410-11. For the third, see *The Born-Einstein Letters*, translated by I. Born (New York: Walker, 1971), p. 148.

72. J. C. Maxwell, *The Scientific Papers of James Clerk Maxwell*, ed. W. D. Niven (Cambridge: University Press, 1890), vol. 1, p. 759.

73. *The Uses of Diversity*, p. 30.

74. *Heretics*, p. 147.

75. For the provenance and context of the Vienna Circle's motto, see my Gifford Lectures, *The Road of Science and the Ways to God* (Chicago: University of Chicago Press, 1978), p. 224 and the preceding half a dozen pages.

76. As advocated in Herbert Marcuse's *One-Dimensional Man*

(1964), which has at least the distinction of having a title perfectly matching its contents.

77. *Heretics*, p. 140.

78. Arnold Lunn, *Now I See* (New York: Sheed and Ward, 1938), p. 51.

79. See John Leo (pseud.), "Cleansing the Mother Tongue: Wanda updates Ralph on orgies, frigidity and one-night stands," *Time* Dec. 27, 1982, p. 78.

80. *Manalive*, p. 169. "Science therefore regards thieves 'in the abstract', just as it regards murderers. It regards them not as sinner to be punished for an arbitrary period, but as patients to be detained and cared for the required period" (ibid., p. 119).

81. And they preach it for a purpose aptly expressed in the *New York Times*'s caption announcing the publication of *Chance and Necessity:* "French Nobel biologist says world based on chance leaves man free to choose his own ethical values" (March 15, 1971, p. 6). The caption was clearly an unwitting echo of some major contentions of Chesterton.

82. I. I. Rabi, "The Interaction of Science and Technology," in *The Impact of Science and Technology,* ed. A. W. Warner and others (New York: Columbia University Press, 1965), p. 20. Such an incomprehension of literature was a piece with Rabi's claim that "science is the only valid underlying knowledge that gives guidance to the whole human adventure [and that] those who are not acquainted with science do not possess the basic human values that are necessary in our time" (quoted in *Time,* May 26, 1967, p. 48). On pondering this statement to which many similar ones could be added from the writings of recent Nobel-laureate scientists and of scientists of slightly lesser rank, one should sense the utter timeliness of Chesterton's antagonism to scientism. Not only the very essence of scientism has remained unchanged since Chesterton's days, but also many of its spots.

83. Jacob Bronowski, *The Origins of Knowledge and Imagination* (New Haven: Yale University Press, 1978), pp. 7-10.

84. *What's Wrong with the World*, p. 152.

CHAPTER THREE: CRITIC OF EVOLUTIONISM

1. Martin Gardner, ed., *Great Essays in Science* (New York: Pocket Books, 1957), p. 77.

2. Nor am I, for that matter an anti-evolutionist, in spite of my criticisms of Darwin and of my sympathy for Chesterton's scathing remarks on integral evolutionism. Since several of my previous publications, such as *The Relevance of Physics* (ch. 7), *The Road of Science and Ways to God* (ch. 18), *Cosmos and Creator* (ch. 4), and *Angels, Apes and Men* (ch. 2) deal in detail with evolution, Darwin and evolutionism, here a brief statement of my views may suffice. I hold the principle of continuity very appealing, most plausible, and scientifically very fruitful. Principles and hypotheses are, however, one thing, proofs and demonstrations another. As to the claim, all too frequently to be found in the "most authoritative" literature, that the Darwinian evolutionary mechanism (the interplay of chance mutations with environmental pressure) has solved all basic problems, I hold it to be absurd and bordering at times on the unconscionable. While the mechanism in question provoked much interesting scientific research, it left unanswered the question of transition among genera, families, orders, classes, and phyla where the absence of transitional forms is as near-complete as ever. As to the origin of life and especially of consciousness, they are today no less irreducible to physics than they were in Darwin's time. I want no part whatever with the position in which Genesis 1-3 is used as a scientific text with predictive value (that is, predicting great lacunae in the fossil record), or with the diametrically opposite

stance in which science is surreptitiously taken for a means of elucidating the utterly metaphysical question of purpose. In short, it is, in my view, intellectually far more honest to keep in mind the grave shortcomings of a theory, however appealing by its unifying and predictive potentialities, than to foster sanguine illusions about its true status, just because one becomes thereby an effectively protected and supported part of the "established consensus."

3. Robert J. Wenke, *Patterns in Prehistory: Mankind's First Three Million Years* (New York: Oxford University Press, 1980, p. 79.

4. *The Everlasting Man*, p. 27. No less forcefully, Chesterton spoke of the Missing Link as one of those "talismanic pictures" whose "power . . . is almost hypnotic to modern humanity" *(What I Saw in America,* p. 188), a statement that would have done credit to any good historian of Darwinism.

5. On Popper and Kuhn as used by creationists, see R. L. Numbers's article, "Creationism in 20th-Century America," *Science* 218 (5 Nov. 1982), 538-44.

6. For details, see ch. 15, "Paradigms or Paradigm," in my *Road of Science*.

7. *Orthodoxy*, pp. 58-59. Nothing would be more mistaken than to conclude from this passage that Chesterton was an "interventionist," that is, postulating special divine creative acts in countless instances in the long evolutionary process. That he never voiced "interventionism" should indicate the largeness of his thinking, which simply left matters open where evidence was unavailable. What he did not consider an open question was the status of things. His Christian orthodoxy, which later wholeheartedly espoused the dogma of transubstantiation, nowhere revealed itself more strikingly in the *Orthodoxy* than in his passionate defense of the reality of things. He rightly sensed that

if Darwinian philosophy was true, "there is no such thing as a thing" (ibid. p. 59), a statement no less profound philosophically than his dictum: "There *is* an Is!" (*St. Thomas Aquinas*, p. 206). Tellingly, only a few pages later (p. 216), Chesterton excoriates the destruction of things by evolutionism!

8. *Lunacy and Letters*, p. 192.

9. *All Is Grist*, pp. 58, 63.

10. *The Everlasting Man*, Prefatory Note. The emphasis given in this chapter to that work should seem justified not only by its excellence, but also by the sheer impossibility of doing justice within the framework of a small book, to the entirety of Chesterton's dicta on man, his favorite topic, as shown by the long columns under that heading in *An Index to Chesterton*.

11. Not all printings of Wells's *Outline of History* carry these statements. In fact, there is a "Third Edition" already printed in 1921 (New York: Macmillan) where the title page merely contains the words, "revised and rearranged by the author."

12. *Fancies versus Fads*, p. 216. Chesterton possibly had in mind his "The Religion of the Reindeer," *New Witness* 15 (6 Feb., 1920) 208-9.

13. As suggested by some lines in a letter of Chesterton to Ronald Knox; see Maisie Ward, *Gilbert Keith Chesterton* (London: Sheed and Ward, 1952), p. 393.

14. Thus Lawrence J. Clipper, *G. K. Chesterton* (New York: Twayne, 1974), p. 115.

15. In the edition quoted above, Jesus was dealt with in less than ten pages, the wars between the Greeks and Persians in over forty.

16. *George Bernard Shaw*, p. 239.

17. *The Everlasting Man*, p. 5. Philosophy is the gist of Chesterton's recall of the genesis of his own thinking on evolution: "My life unfolded itself in the epoch of evolution, which

really means unfolding. But many of the evolutionists of that epoch [1880–1900] really seemed to mean by evolution the unfolding of what is not there. I have since, in a special sense, come to believe in development; which means the unfolding of what is there" *(Autobiography,* pp. 51–52).

18. See *The Autobiography of Charles Darwin, 1809–1882,* with original omissions restored, edited with appendix and notes by his granddaughter, Nora Barlow (New York: W. W. Norton, 1969), pp. 107–8.

19. See Darwin's letter of March 29, 1863, to J. D. Hooker, in Francis Darwin, *The Life and Letters of Charles Darwin* (London: 1887), vol. 3, p. 18.

20. And a piece with Darwin's long-standing materialism amply evidenced by his early Notebooks; for details, see my *Angels, Apes and Men,* pp. 52–53.

21. In fact, according to him, his colleagues "too often swept under the carpet the biggest problem in biology, the existence of consciousness," See *Supplement to Royal Society News,* Issue 12, Nov. 1981, p. v.

22. *The Everlasting Man,* p. 6

23. The reason for my singling out the folio work, *La caverne du Font de Gaume* (Monaco: A. Chene, 1912), by L. Capitan, H. Breuil, and D. Peyrony, is that its wealth of illustrations was a chief source for E. A. Parkyn's *An Introduction to the Study of Prehistoric Art* (London: Longmans, Green & Co., 1915), possibly the best and certainly the most accessible book on the subject for the English public during the decade preceding *The Everlasting Man.* In 1940, the Abbé Breuil was the fourth to enter the caves of Lascaux, first briefly explored by two boys.

24. *The Everlasting Man,* p. 10.

25. Ibid., p. 16. It is well to recall that the existence of so many beautiful patterns in the realm of living was considered by Darwin "fatal" to his theory not only if it could be proven

that beauty existed for the delight of man or of the Creator but also if it could be proven that beauty arose "for the sake of mere variety" and not strictly for usefulness. This latter point should seem all the more important as Darwin fully admitted that "many structures [noted for their beauty] are now of no direct use to their progenitors, and may never have been of any use to their progenitors." *The Origin of Species* (6th ed.: London: John Murray, 1876), p. 160. In the same breath Darwin asked his critics to prove their case as if the chief burden of proof had not been on him!

26. *The Everlasting Man*, p. 18.

27. *The Thing*, p. 100.

28. *The Everlasting Man*, pp. 17, 22.

29. John E. Pfeiffer, *The Creative Explosion: An Inquiry into the Origins of Art and Religion* (New York: Harper & Row, 1982). In this book a careful study is made of all the additional data on paleolithic art that accrued since the death of the Abbé Breuil (1961), none of which make that art less human and less explosively sudden.

30. *The Everlasting Man*, p.19.

31. Ibid., pp. 16–19.

32. Ibid. p. 26.

33. The evidence published in mid-December 1982 (see *Nature* 300 [1982]: 631–35, and the *New York Times,* Dec. 16, 1982, p. B16) was only the beginning of even greater revisions. On May 3, 1983, Sandra Blakeslee reported in the *New York Times* (p. C1): "The debate over whether the primate Lucy actually stood up on two feet three million years ago and walked—thus becoming one of mankind's most important ancestors—has evolved into two interpretative viewpoints, three family trees, spats over four scientific techniques, and too many personality clashes."

34. *The Everlasting Man*, p. 29.

35. *Fancies versus Fads,* p. 214. Those ready to cry at this point, *lèse majesté,* are kindly invited to ponder a phrase of Darwin written shortly after the publication of the second edition of the *Origin:* "I believe in nat. selection not because I can prove in any single case that it has changed one species into another, but because it groups and explains well (as it seems to me) a lot of facts in classification, embriology, morphology, rudimentary organs, geological succession & distribution." A facsimile of that letter, catalogued in the British Museum as A DD Ms 37725 f6, faces the title page of *L'évolution du monde vivant* by Maurice Vernet (Paris: Plon, 1950). If it is wise to treat with caution theories which are supported by some obvious, though far from overwhelming experimental evidence, it should seem rather unwise, if not outright foolish, to venerate as final truth a theory which lacks such support even according to its very architect.

36. *Come to Think of It,* p. 149.

37. Revealingly, there was little if any hint in Attenborough's recital as to the complete inability of Darwinian theory, or any evolutionary and biochemical theory, to explain the emergence of any of those countless marvelous devices, which keep revealing themselves in the measure in which research progresses. A recent example is the spotting of pure single crystals of the iron oxide known as magnetite in viruses as their means of orientation in the earth's magnetic field (see *Nature* 302 [March 31, 1983], p. 411). That such devices defy Darwinian theory has, of course, been occasionally admitted by Darwinists, though fleetingly enough so as not to make too much impact. A case in point is Stephen Jay Gould's "Opus 100," published in the April 1983 issue of *Natural History* (pp. 10-21). It is a report of his investigations of the Bahamian land snail, *Cerion,* or rather of its extraordinary varieties. Gould would have been worth quoting

even for his introductory aside, in which he justified research into minutiae "when all the giddy generalities of evolutionary theory beg for study in a lifetime too short to manage but a few." The justification specified by him lies, of course, in the fact that investigation of the smallest details brings out the fundamental and overall problems. For precisely in succeeding in his task as a taxonomist is the Darwinian naturalist brought face to face to the problem of finding a mechanism that would produce a very large number of variations (species) on a common stock. The likelihood that "a complex set of independent traits can evolve in virtually the same way more than once" is of no help according to Gould, who admits: "I do not see how this can happen if each must develop separately, following its own genetic pathway, each time." After speculating about some master switch (with no master of course), Gould continues: "*Cerion* provides insight into what may be the most difficult and important problem in evolutionary theory: How can new and complex forms (not merely single features of obvious adaptive benefit) arise if each requires thousands of separate changes, and if intermediate stages make little sense as functioning organisms. . . . Yet so deep is our present ignorance about the nature of development and embryology that we must look at final products . . . to make uncertain inferences about underlying mechanisms [such as master switches]." Those familiar with the literature since Darwin, will easily recognize in this admission a long-standing pattern, destructive of the view that ever since Darwin his evolutionism has been accepted by the overwhelming majority of biologists. A careful investigation by Peter J. Bowler *(The Eclipse of Darwinism: Anti-Darwinian Evolution Theories in the Decades around 1900* [Baltimore: Johns Hopkins University Press, 1983]) of the record for those forty or so years revealed the persistence of so many conspicuous dissenters as to prompt him to deny for that period to Darwinism the status of being a paradigm, that is, an essentially undisputed scientific conviction. A similar investigation of the record for the past six decades may easily lead to much the same conclusion. See also my review of Bowler's book in *The Tablet* (London) Feb. 11, 1984, pp. 135–36.

38. *All I Survey*, p. 230.

39. *Encyclopédie Française, Tome 5: Les êtres vivants* (Paris: Société de Gestion de l'Encyclopédie Française, 1937), p. 82.

40. *Fancies versus Fads*, pp. 216–23.

41. In his *Man's Place among the Mammals* (New York: Longmans Green & Co., 1929) Frederick Wood Jones (1879–1954), then professor of physical anthropology at the University of Hawaii and later professor of human and comparative anatomy in the Royal College of Surgeons (London) and all the time a firm believer in the exclusivity of strictly natural causes in scientific explanations, called for a counterrevolution in words that would have done credit to the most articulate among present-day paradigmists: "If, after a painful struggle and a deal of emotional and sentimental opposition, some revolutionary thesis is accepted by the bulk of more or less thinking people, there is even a tendency to regard the matter as finally settled. Opinion has been rudely upset, sentiment has been outraged, emotions deeply stirred; but once the change of thought has been effected there is a dislike for facing a second disturbance of belief—a reluctance to retreat from a position that was only accepted after so painful a struggle. Yet reverence for fact is the only attitude compatible with the scientific spirit, and at times retractions and disavowals are absolutely necessary" (pp. 2–3). A year before Chesterton died, Earnest Albert Hooton, professor of anthropology at Harvard, concluded his massive *Up from the Ape* (New York: Macmillan, 1935) with remarks with which Chesterton could have wholly agreed: "However you look at him, man is a miracle, whether he be a miracle of chance, or

nature, or of God. . . . That evolution has occurred I have not the slightest doubt. That it is an accidental or chance occurrence I do not believe, although chance probably has often intervened and is an important contributing factor. But if evolution is not mainly a chance process it must be an intelligent or purposeful process." While Chesterton would have taken a vehement exception to Hooton's question, "What difference does it make whether God is Nature or Nature is God?", his reasons would have been a probing into Hooton's ensuing words: "The pursuit of natural causes either leads to the deification of Nature, or to the recognition of the supernatural, or to a simple admission of ignorance, bewilderment, and awe" (p. 604). Many other authors, prominent in those decades, could also have been quoted, if space permitted, to the same effect.

42. The statement is part of the review by James Gray, professor of zoology at Cambridge, of Julian Huxley's *Evolution in Action* in *Nature* 173 (1954): 227.

43. *The Thing*, p. 53. Almost twenty years after this statement was made by Chesterton, Professor Frederick Wood Jones wrote: "Missing Links have been discovered with quite commendable regularity. All have been acclaimed in their heyday and by their partisans as settling the whole business, but the status of most of these discoveries has had to undergo considerable readjustment with the passage of time and with the more detached and critical methods applied to their examination. Some have suffered the fate of being relegated to an agreed oblivion from which it is not considered scientifically discreet to resuscitate them." *Hallmarks of Mankind* (Baltimore: Williams, 1948), p.3. Two years later, 1950, in surveying the latest evidence from South Africa, of which he was a part discoverer, Robert Broom remarked in a way of conclusion which would have fully satisfied Chesterton with its scientific honesty: "Though we can say with complete confidence that man has evolved from an Old World primate,

and can even come very near to the ancestor that gave rise to the human line, we cannot say what has brought about the evolution" *Finding the Missing Link* (London: Watts & Co., 1950), p. 91.

44. *The Thing*, p. 89. The immediate issue was Keith's unfamiliarity with the work of Louis Vialleton, professor of comparative anatomy at the University of Montpellier, whose great books, *Membres et ceintures des vertèbres tetrapodes* (1924) and *L'origine des êtres vivants* (1929), have largely been ignored in the Anglo-Saxon world. The same is true of *Grundfragen der Palaeontologie* (1950) by Otto H. Schindewolf, professor of geology and paleontology at the University of Tubingen.

45. Such was at least the thinly disguised thrust of Keith's Presidential Address, "Darwin's Theory of Man's Descent as It Stands Today," at the meeting of the British Association for the Advancement of Science, in Leeds, 1927.

46. A fact noted with undisguised satisfaction by Howard E. Gruber, author of *Darwin on Man: A Psychological Study of Scientific Creativity,* a work published together with *Darwin's Early and Unpublished Notebooks,* transcribed and annotated by Paul H. Barrett, with a foreword by Jean Piaget (New York: E. P. Dutton, 1974). The notebooks date from 1837-39.

47. *The Thing*, pp. 90-91.

48. No less a Darwinist than Gaylord G. Simpson admitted: "It is a feature of the known fossil record that most taxa appear abruptly. . . . Gaps among known orders, classes, and phyla are systematic and almost always large." See his "The History of Life" in Sol Tax, ed., *The Evolution of Life,* vol. I (Chicago: University of Chicago Press, 1960), p. 149.

49. David L. Hull, "The Metaphysics of Evolution," *British Journal for the History of Science* 3 (1967): 309-37; for quotation see p. 337. The extent to which the notion of species as a

universal has confronted biologists—from Lamarck through Darwin and Spencer to the present-day spokesmen of molecular biology—with perplexities whose genuine metaphysical status they were all too often unable to recognize, is the topic of the octogenarian Etienne Gilson's masterpiece, *D'Aristôte à Darwin et retour* (1971), now available in John Lyon's English translation, *From Aristotle to Darwin and Back Again* (Notre Dame, Ind.: University of Notre Dame Press, 1984) with my introduction.

50. *The Everlasting Man*, p. xvii.

51. *Lunacy and Letters*, pp. 78–79.

52. *Avowals and Denials*, pp. 23–24.

53. *The Ball and the Cross*, p. 1.

54. *What's Wrong with the World*, pp. 268–69.

55. *Orthodoxy*, pp. 204–5.

56. *The Spice of Life*, p. 118.

57. *Lunacy and Letters*, p. 118.

58. *Charles Dickens*, p. 275.

59. *All Is Grist*, p. 117.

60. *Heretics*, pp. 76, 171.

61. George Bernard Shaw, *Heartbreak House* (London: Constable & Co., 1919), p. xiii.

62. *The Ball and the Cross*, p. 279. On Chesterton's acceptance of the divinity of Christ as early as 1903 and with a full realization of the bearing of such belief on the study of man, see note 51 to ch. 1. It was Chesterton's perception of the paramount importance of the soul that made him see the basic mistake of scientistic Utopians, like H. G. Wells, in their denial of it. See *Heretics*, p. 79.

63. Throughout ch. 3 of *The Descent of Man* Darwin argued that there were only differences of degree, not of kind, between animals and men with respect to *any* intellectual, moral, and social faculty.

64. For the text of Philip Morrison's Jacob Bronowski Lecture,

"Termites and Telescopes," see *The Listener*, Aug. 23, 1979, pp. 234–38.

65. One of those Darwinists, Theodore Dobzhansky, certainly could not derive from his lecture on human evolution, given in 1969 at my university, Seton Hall, the tenet of the essential equality of all men and races, when confronted by a visibly upset black student in the question-answer period.

66. See, for instance, Darwin's letter of July 3, 1881, to W. Graham in Francis Darwin, *The Life and Letters of Charles Darwin*, vol. 3, p. 316. The Chestertonian empirical argument about the everlastingness of man would equally bear on the inability of consistent Darwinists to see anything revolting in the estimated fifty million or so abortions now performed clinically all over the world every year. The same argument should also serve as the proper backdrop when faced with such hideous news as the advertisement for facial cream, called California Beauty, produced by René Ibry, Inc., of Cannes, France: "Exclusively taken from fetuses, the young cells applied to old tissues are able to regenerate the latter. These cells are all the more effective because they are living." Quoted by Grace O. Dermody, "Why Thousands Will March to End Abortion," in *New York Times*, Jan. 16, 1983, p. NJ24.

67. *What I Saw in America*, pp. 301–2.

68. *Fancies versus Fads*, p. 223.

69. For a detailed report on this controversy, aired at a symposium organized by the New York Academy of Sciences, see *New York Times*, Feb. 22, 1983, p. C1, "Did Flight Begin on the Ground?" by B. Webster.

70. Ibid. p. C5.

71. *Fancies versus Fads*, p. 222.

72. For an excellent survey of the modifications in the concept of natural selection during the first hundred years after the

publication of Darwin's *Origin*, see Philip G. Fothergill, *Evolution and Christians* (London: Longmans, 1961), pp. 199-227.

73. T. H. Huxley, "Biogenesis and Abiogenesis" (1870), in *Discourses: Biological and Geological* (London: Macmillan, 1894), pp. 256-57.

CHAPTER FOUR: CHAMPION OF THE UNIVERSE

1. As discussed in Chapter One.

2. *Lunacy and Letters*, p. 123.

3. Ian Boyd , *The Novels of G. K. Chesterton*, p. 240.

4. *The Poet and the Lunatics*, p. 125.

5. Maisie Ward, *Gilbert Keith Chesterton* (London: Sheed and Ward, 1944), pp. 48, 58, 59, 64.

6. This is a point all too often far better perceived by the antagonists of theism than by its supporters. The *Autobiography* of J. S. Mill and *The Education of Henry Adams* contain in this respect particularly revealing passages and so do the works of Spencer and Engels, to refer only to some noted authors pertaining to the period in question.

7. The *Daily News* essay fills a column on p. 8. Its photocopy was obtained for me by Dr. Peter E. Hodgson, Fellow of Corpus Christi College, Oxford. Gary Wills quoted a few lines of the essay in his *Chesterton* and Alan L. Maycock reprinted about a quarter of it in *The Man Who Was Orthodox: A Selection from the Uncollected Writings of G. K. Chesterton* (London: Dennis Dobson, 1963), p. 177, under the caption "Divine Immanence."

8. *The Defendant*, p. 48.

9. Chesterton did so partly in the pages of *Clarion*, a weekly owned and edited by Blatchford (1851-1943), where there appeared first in installments Blatchford's *God and My Neighbour*, an immensely popular manifesto of atheism, which ultimately

sold in a million or so copies. Blatchford would for months reserve the pages of *Clarion* exclusively for his critics, among them Chesterton, who characterized his gesture as one of "unparalleled generosity." See Laurence Thompson, *Robert Blatchford: Portrait of an Englishman* (London: Victor Gollancz, 1951), p. 168. In a letter to Thompson, Chesterton paid the following tribute to Blatchford: "Very few intellectual swords have left such a mark on our time, have cut so deep, or remained so clean. . . . Blatchford was an artist if ever there was one in this world. His triumphs were of strong style, native pathos and picturesque metaphor; his very lucidity was a generous sympathy with simpler minds. For the rest, he has triumphed by being honest and by not being afraid." The reference to "simpler minds" is noteworthy. Lucidity and honesty did not necessarily mean depth. Or as Chesterton also warned: "Blatchford is a very honest man, only unfortunately he is entirely divorced from the realities of human life" (ibid., pp. 172, 216). What this implied for Blatchford's grasp of science, may be gathered from note 12 below. See also note 51 to Chapter One.

10. L. Hunter, *G. K. Chesterton: Explorations in Allegory* (New York: St. Martin's Press, 1979), p. 41.

11. *Heretics*, p. 195.

12. Blatchford was well-meaning in more than one sense. As all too often is the case, Blatchord's atheism too was cover-up for mystical longings which only waited for the proper opportunity to assert themselves in full. Not that Blatchford would have admitted that the death of his wife in 1921 was far more decisive in turning him into a devotee of spiritism than was his reading about the electronic structure of atoms in the books of Oliver Lodge, a physicist and an outspoken spiritist following the death of his son. It was on that bit of science that Blatchford based his inference: "If a bullet is composed of electric charges, may

not a human spirit be composed of something equally intangible—or tangible?" See Blatchford, *My Eighty Years* (London: Cassell, 1931), p. 263. Blatchford could indeed be so naively "mystical" as to not perceive the irony of his dictum: "A Determinist must live up to himself." See Thompson, *Robert Blatchford*, p. 173.

13. See Maycock, *The Man Who Was Orthodox*, pp. 89-90.

14. *Heretics*, pp. 12, 15-16, 305.

15. *Orthodoxy*, pp. 14, 58.

16. A patriot who held high, of course, "The Flag of the World," the theme of ch. 5 of *Orthodoxy*.

17. See, for instance, Pierre Duhem, *The Aim and Structure of Physical Theory*, trans. P. P. Wiener (1954; New York: Atheneum, 1962), pp. 287-88.

18. See Karl R. Popper, *Conjectures and Refutations* (New York: Harper & Row, 1968), p. 136.

19. Popper is not only shy but also very skeptical, as shown by his remarks on the theory of the formation of elements as supported by the cosmic background radiation. No less revealing is the eagerness with which he refers to Jean-Paul Vigier's now fairly old and not-at-all convincing arguments against the recessional red-shift as an indication of the expansion of the universe. See K. R. Popper, *The Open Universe: An Argument for Indeterminism* (Totowa, N.J.: Rowman and Littlefield, 1982), pp. 142-43.

20. As is very clear from scientific efforts aimed at deriving the universe from a mere quantum flip or simply from nothing and from the speculations of logicians, such as Rudolf Carnap, whose "logical build-up of the world" *(Der logische Aufbau der Welt,* 1928) is markedly void of references to the universe.

21. *Orthodoxy*, p. 69.

22. *St. Thomas Aquinas*, pp. 193-95. Chesterton's chastizing there the modern critic who wants to meet the Schoolman as

his intellectual equal, without first taking a note of "what the medieval Schoolman meant by form" (p. 195), would, of course, be relevant to the "critical," that is, "transcendental" Thomists.

23. *Orthodoxy*, p. 99.

24. *The Glass Walking Stick and Other Essays*, pp. 1–4.

25. *Orthodoxy*, p. 115.

26. As argued by Professor William H. McCrea during a conference on "Cosmology, History and Theology" at the University of Denver, November 5–8, 1974. He was considerably taken aback by my question, whether the wall facing him was also a mere sensation on his retina.

27. Aristotle, *Metaphysics*, 1074b–75b. Such was both the capstone and very source of that cosmic necessitarianism which put science into a straitjacket for almost two thousand years; that is, until some basic points of Aristotelian cosmology were rejected by medieval Scholastics, guided by the dictates of their Christian faith in creation out of nothing and in time.

28. This is the argument of ch. 5, "The Flag of the World," in *Orthodoxy*.

29. See my articles, "From Scientific Cosmology to a Created Universe," in *The Irish Astronomical Journal* 15 (March 1982): 253–62, and "Das Gravitations-Paradoxon des unendlichen Universums," in *Sudhoffs Archiv* 63 (1979): 105–21.

30. That consensus is the topic of ch. 8, "The Myth of One Island," in my book *The Milky Way: An Elusive Road for Science* (New York: Science History Publications, 1972).

31. For a survey of those theories, see ch. 6, "The Angular Barrier," in my *Planets and Planetarians: A History of Theories of the Origin of Planetary Systems* (Edinburgh: Scottish Academic Press; New York: J. Wiley, 1978).

32. A passage already quoted in full; see note 11 above.

33. H. G. Wells, *First and Last Things: A Confession of Faith and Rule of Life* (London: Watts, n.d.), p. 30.

34. *The Well and the Shallows*, p. 77. Chesterton may have just as well have referred to Jeans's *The Mysterious Universe*, possibly the most widely read high-level popularization of science published in modern times.

35. *St. Thomas Aquinas*, pp. 215-16.

36. *The Defendant*, p. 58.

37. *Orthodoxy*, pp. 112, 143-44. Clearly, Chesterton did not plead for a physically small universe.

38. *G. F. Watts*, p. 35.

39. *The Resurrection of Rome*, p. 264.

40. Contrary to a suggestion of Lawrence Clipper, who in his *G. K. Chesterton* (New York: Twayne, 1974), p. 90, quoted in support the foregoing statement of Chesterton. In *The Resurrection of Rome* Thomas Aquinas is mentioned a few times but for very different reasons. In the context Chesterton explains himself in a genuinely Thomistic sense, which goes far deeper than mere size. According Chesterton, a nearby object, be it as small as a pearl, can far more effectively activate man's sensitivity for the existence of *things* than can large objects, such as the sky, which, because of their remoteness, may easily fade into a mere *theory* (see p. 265). The philosophical implication of Chesterton's fondness for things can best be gathered from an autobiographical reminiscence of Gilson about his first exposure to idealist metaphysics, an experience much antedating his discovery of Thomism: "I understood well enough the meaning of each and every particular sentence, but I failed to grasp the ultimate purpose of the discourse. Unawares, I was already plagued with the incurable metaphysical disease they call *chosisme,* that is, crass realism. Today there is no intellectual infirmity more utterly

despised, but I know well enough that one cannot get rid of it. A person like myself who suffers from this ailment speaks only about 'things' or about propositions related to things. I was as nonplused by my first contact with idealism as I was later to be by my first encounter with the so-called philosophies of the mind" *(The Philosopher and Theology,* translated from the French by Cecile Gilson [New York: Random House, 1962], p. 18). Had Gilson been born forty or so years later, he would have been no less nonplused by an initiation to philosophy through "scientific" philosophies or simply through much of what passes nowadays for philosophies of science.

41. *St. Thomas Aquinas,* pp. 204, 206.

42. Ibid. When such a series of extraordinarily forceful philosophical phrases comes from the pen of one who is disdainfully spoken of by some "realist" philosophers of science as a mere journalist, one wonders whether the latter would not perform far better in the prestigious chairs of the former.

43. *The Man Who Was Thursday,* p. 13.

44. "Boswell" (1903); in *G.K.C as M.C.,* p. 3.

45. In more than one sense. Soundly admirable should seem the feverish speed with which the papers presented at the study week on cosmology and fundamental physics at the Pontifical Academy of Sciences (September 28-October 2, 1981) were published *(Astrophysical Cosmology,* ed. H. A. Brück *et.al.* [Vatican City: Pontificia Academia Scientiarum, 1982]), lest they should appear out of date by the time of publication. Hardly so sound should seem some extravagant speculations (and their consideration with equanimity) in which, as for instance is the case with the so-called inflationary theory of the beginning of the universe, transition from non-being is taken for a "normal" process.

46. *Charles Dickens,* p. 289.

47. For details, see *The First Three Minutes: A Modern View*

of the Origin of the Universe (London: André Deutsch, 1977), by Steven Weinberg, easily the best popularization of the scientific side of the origin of the universe. Some of Weinberg's philosophical statements are flippant, to say the least.

48. For details and data, see the very readable article of B. J. Carr, "On the Origin, Evolution and Purpose of the Physical Universe," *Irish Astronomical Journal* 15 (March 1983): 237-53. Like Weinberg's book, this article too is tainted with philosophical flippancies.

49. *Orthodoxy*, p. 114.

50. Clipper, *G. K. Chesterton*, p. 99.

51. *Heretics*, p. 43.

52. *All Is Grist*, pp. 7-8. Not surprisingly, Chesterton equated shapelessness with evil, nay with Satan himself. See *A Miscellany of Men*, p. 146.

53. It has Carr (see note 48 above) as one of its learned and ardent proponents.

54. He would refer to ch. 1 of *Heretics* and to various passages in *Orthodoxy*, such as pp. 113 and 145.

55. *Fancies versus Fads*, p. 143.

56. Arthur S. Eddington, *New Pathways in Science* (Cambridge: University Press, 1934), p. 217.

57. *Robert Browning*, p. 218.

58. *Fancies versus Fads*, p. 91.

59. *Orthodoxy*, p. 275.

60. *As I Was Saying*, p. 28.

61. As argued in ch. 17, "Cosmic Singularity," in my Gifford Lectures, *The Road of Science and the Ways to God*.

62. For a summary and criticism of such theories, see ibid.

63. *Orthodoxy*, p. 139.

64. *A Miscellany of Men*, p. 262.

65. *All Is Grist*, p. 148. For Dawson's reference to Duhem's

Système du monde, of which only the first five volumes were in print in 1928, see his *Progress and Religion: An Historical Enquiry* (London: Sheed and Ward, 1929), p. 143. The publication of the last five, and in that respect even more significant, volumes of the *Système,* was delayed for a reason which would not have surprised Chesterton and which is now set forth on the basis of hitherto unpublished material in my article, "Science and Censorship: Hélène Duhem and the Publication of the *Système du monde,*" *Intercollegiate Review,* 20 (Winter 1985), pp. 41–49.

66. *All Is Grist,* p. 150.

67. *The Resurrection of Rome,* p. 120.

68. For texts and background, see my *Science and Creation: From Eternal Cycles to an Oscillating Universe* (Edinburgh: Scottish Academic Press; New York: Science History Publications, 1974), pp. 179–82.

69. *The Resurrection of Rome,* p. 121.

70. "The Book of Job," in *G.K.C. as M.C.,* p. 42.

71. *The Everlasting Man,* p. 25.

72. "The Book of Job," p. 42.

73. *George Bernard Shaw,* pp. 241–42.

74. More specifically, Kant wanted to provide a strictly intellectual foundation for Rousseau's sentimental program. For details and documentation, see my *Angels, Apes and Men* (La Salle, Ill: Sherwood Sugden, 1983), pp. 27–33.

75. *Orthodoxy,* p. 58.

76. *A Miscellany of Men,* p. 114.

77. *Heretics,* p. 290.

Index

Note on the Author

Stanley L. Jaki, a Hungarian-born Catholic priest of the Benedictine order, is Distinguished Professor at Seton Hall University, South Orange, New Jersey. With doctorates in theology and physics, he has for the past twenty-five years specialized in the history and philosophy of science. The author of sixteen books and over seventy articles, he served as Gifford Lecturer at the University of Edinburgh and as Fremantle Lecturer at Balliol College, Oxford. He is recipient of the Lecomte du Nouy Prize and has lectured at major universities in the United States, Europe, and Australia. He has recently been elected *membre correspondant* of the Académie Nationale des Sciences, Belles-Lettres et Arts of Bordeaux.